Fundamental Physics
by
Trevor G. Underwood

By the same author:

The Surface Temperature of the Earth. (November 2019);

*Urbain Le Verrier on the Movement of Mercury
- annotated translations.* (November 2021);

*Quantum Electrodynamics – annotated sources.
Volumes I and II.* (April 2023);

Special Relativity. (June 2023);

General Relativity. (November 2023);

Gravity. (March 2024);

Electricity & Magnetism. (May 2024);

Quantum Entanglement. (June 2024);

The Standard Model. (September 2024);

New Physics. (October 2024);

The Cosmological Redshift of Light. (November 2024);

Cosmic Microwave Background Radiation. (January 2025).

all distributed by Lulu.com.

Published by Trevor G. Underwood
18 SE 10th Ave.
Fort Lauderdale, FL 33301

ISBN: 979-8-218-69757-0 (hardcover)
Library of Congress Control Number: 2025911049

Printed and distributed by Lulu Press, Inc.

700 Park Offices Dr.
Ste. 250
Durham, NC, 27713
http://www.lulu.com/shop

PREFACE

This volume provides a summary of the current state of fundamental physics by bringing together the prefaces, which provide a summary of each of the eleven volumes that I published between March 2023 and January 2025.

Examination of new papers which attempt to integrate the fourth interaction, *gravity*, with the Standard Model of particle physics, describing the other three fundamental forces of nature, the *electromagnetic*, *weak* and *strong interactions*, demonstrated the need for a single volume which pulls together the current state of all of the theories described in my books.

My conclusion in "*New Physics*" was that "e*lementary* and *composite* particles with *mass* attract each other through the *gravitational interaction or gravitational force* resulting from *quantum entanglement between matter*", which introduced "a *quantum theory of gravity*". But it was not clear how and to what extent the new theories corresponded to this. So, my first task was to summarize my previous analysis on which this conclusion was based.

CONTENTS

Quantum Electrodynamics – an annotated sourcebook.

Quantum physics is the study of matter and energy at the most fundamental level.

Quantum mechanics is the branch of physics that deals with the behavior of matter and light on a subatomic and atomic level.

Quantum electrodynamics deals with the electromagnetic field and its interaction with electrically charged particles.

Quantum field theory is relativistic quantum mechanics, the synthesis of quantum mechanics with special relativity.

I had decided that in my self-imposed lockdown to avoid COVID, I could usefully pick up where I had left off in 1965, after graduating in theoretical physics from Cambridge University. As the volume of papers relevant to understanding the current state of *quantum electrodynamics* was extremely large and some were not easily obtainable, and a few had not been translated into English, I decided to produce an annotated sourcebook from which it would be easier to understand the current state of our knowledge about the electromagnetic field and its interaction with electrically charged particles. This would not have been possible in 1965, before the advent of personal computers in 1971, the origin of internet in 1969, the increase in disc capacity, and the availability of much of this material on the internet in the last two decades.

These two volumes were spun off from a larger project involving a review of the current state of our scientific understanding of electromagnetic radiation and gravity. This is not a sourcebook in the conventional sense. It is a working document which brings together annotated extracts from 107 primary sources, or translations of them, of the development of quantum electrodynamics so that it is easier for a researcher to deal with the large volume of relevant sources. Papers hidden behind pay walls for which no alternative internet source or abstract is available have been omitted, but fortunately these are of little consequence. Links to internet copies of the primary material or alternative sources are provided where available to enable these to be consulted. A summary is provided at the head of each paper and in the Contents. The references in each paper are expanded to include the title (and its translation where relevant), and a copy of the summary where this is available and helpful to avoid unnecessary cross referencing. Biographies of the main contributors are also provided, of which 18 received Nobel Prizes in Physics. Explanations of terminology (which is highlighted in italics) and biographical information are largely drawn from Expedia, unless otherwise referenced. The chronological development of the theories of quantum mechanics and quantum electrodynamics constituted an interesting interplay between theory and experiment.

Volume I, covering the period from 1896 to 1931 is primarily focused on the development of the largely successful *non-relativistic* theory of quantum mechanics and quantum electrodynamics. Volume II, covering the period from 1930 up until 1965, when Sin-Itiro

Tomonaga, Richard Feynman, and Julian Schwinger received their Nobel prizes, addresses the attempts to formulate a *relativistic* quantum electrodynamics or quantum field theory for the electron when the energy of the electron is *relativistic*, and in particular to address, through a process of *renormalization*, the still unresolved divergencies arising largely, if not entirely, from the assumption of a point electron.

Despite the claims to the contrary in modern textbooks, there have been no significant developments in the quantum electrodynamics or quantum field theory since 1965 to resolve the underlying occurrence of divergencies recognized by Paul Dirac, Tomonaga, Schwinger and Feynman. Witness to this is the very comprehensive 2018 *An Introduction to Quantum Field Theory* by Michael Peskin and Daniel Schroeder, first published in 1995, which replaced the 1965 two-volume text by James Bjorken and Sidney Drell, *Relativistic Quantum Fields*, which focuses on the application of Feynman diagrams. The former claims that "Quantum Electrodynamics (QED) is perhaps the best fundamental physical theory we have" then devotes Part II (pages 265-470), nearly one third of the book, to *renormalization*.

Born, in his 1954 Nobel prize speech, noted that "Planck, himself, belonged to the sceptics until he died. Einstein, De Broglie, and Schrodinger have unceasingly stressed the unsatisfactory features of quantum mechanics and called for a return to the concepts of classical, Newtonian physics while proposing ways in which this could be done without contradicting experimental facts".

Schwinger, in the Preface of his 1958 book [*Selected Papers on Quantum electrodynamics.*], "questioned whether *renormalization* simply corrected a mathematical error that causes the divergencies, or whether *there is a serious flaw in the structure of field theory*". He concluded that "the observational basis of quantum electrodynamics is self-contradictory" and that "a convergent theory cannot be formulated consistently within the framework of present space-time concepts" … "It can never explain the observed value of the dimensionless coupling constant measuring the electron charge … a full understanding of the electron charge can exist only when the theory of elementary particles has come to a stage of perfection that is presently unimaginable".

Tomonaga, in his 1965 Nobel prize speech, note that "In order to overcome the difficulty of an infinitely large *electromagnetic mass, Lorentz considered the electron not to be point-like but to have a finite size. It is very difficult, however, to incorporate a finite sized electron into the framework of relativistic quantum theory.* Many people tried various means to overcome this problem of infinite quantities, but nobody succeeded".

Feynman, in his 1965 Nobel prize speech, described *renormalization* as "simply a way to sweep the difficulties of the divergences of electrodynamics under the rug".

Dirac's final judgment on *quantum field theory*, in his last paper published in 1987 [*The inadequacies of quantum field theory.*], was that "These rules of *renormalization* give surprisingly, excessively good agreement with experiments. Most physicists say that these working rules are, therefore, correct. I feel that is not an adequate reason. Just because the results happen to be in agreement with observation does not prove that one's theory is correct."

I was tempted to include my own initial analysis of the problem but decided against this so that others were free to draw their own conclusions and spot the error or errors, and, ideally, provide a solution free of the divergencies, for which they might be awarded a Nobel prize.

1   Quantum Electrodynamics. Volume I

The move away from Maxwell's electromagnetic wave theory began in 1892, when Hendrik Lorentz presented his electron theory, for which together with Pieter Zeeman he received a half share of the 1902 Nobel Prize. This posited that in matter there are charged particles, electrons, that conduct electric current and whose oscillations give rise to light. In March 1897, at the age of 32, Zeeman published a paper [*On the influence of magnetism on the nature of the light emitted by a substance.*] which examined how light was affected by magnetic fields, which Faraday had attempted in 1862 but without success. Using a more powerful instrument Zeeman showed that, under the influence of a magnetic field, the lines in a spectrum split up into several lines, known as the *Zeeman effect*. This phenomenon could be explained by the electron theory formulated by Lorentz. [Lorentz, H.A. (1892). *La Théorie électromagnétique de Maxwell.* and (1895). *Versuch einer Theorie der electrischen und optischen Erscheinungen in bewegten Körpern.*]

In April 1898, Thomas Preston published a paper [*Radiation Phenomena in the Magnetic Field.*] describing further fine structure in the separation of spectral lines in a strong magnetic field, subsequently described as the *Anomalous Zeeman Effect*. This was not explained by the current theory of the atom based on charged electrons rotating about a nucleus.

In 1900, Max Planck derived the first version of the famous Planck *black-body radiation* law, which described the experimentally observed black-body spectrum well. *This first derivation did not include energy quantization*, and did not use statistical mechanics. Planck then revised his first approach, relying on Boltzmann's statistical interpretation of the second law of thermodynamics as a way of gaining a more fundamental understanding of the principles behind his radiation law. The central assumption behind his new derivation presented in December 1900 [Planck, M. (1901). *Ueber das Gesetz der Energieverteilung im Normalspektrum.* (On the Law of Distribution of Energy in the Normal Spectrum.)] was the supposition, now known as the Planck postulate, that *electromagnetic energy could be emitted only in quantized form, in other words, the energy could only be a multiple of an elementary unit $E = hv$*, where h is Planck's constant and v is the frequency of the radiation. From this Planck arrived at the famous formula for the density of a black body radiation at temperature T. This marked the beginning of the quantum theory of radiation.

In a paper published in 1913 [*On the Constitution of Atoms and Molecules, Part I.*], Neils Bohr, at the age of 28, took the next step by adapting Rutherford's theory of atomic structure to Planck's quantum hypothesis creating his model of the atom. He supposed the atom to consist of a nucleus with a positive charge and electrons with charge a negative charge, moving according to the laws of classical mechanics. He introduced the idea that an electron could drop from a higher-energy orbit to a lower one, in the process emitting a quantum of discrete energy. This became a basis for what is now known as the *old quantum theory*. From a set of assumptions concerning the stationary state of an atom and the frequency of the radiation emitted or absorbed when the atom passes from one such state to another, he showed that it is possible to obtain a simple interpretation of the main laws governing the line spectra of the elements.

In a paper published in 1914 [*Beobachtungen über den Effekt des elektrischen Feldes auf Spektrallinien I. Quereffekt*. (Observations of the effect of the electric field on spectral lines I. Transverse effect)], Johannes Stark described the shifting and splitting of spectral lines of atoms and molecules due to the presence of an external electric field, subsequently known as the *Stark effect*, the electric-field analogue of the *Zeeman effect*.

Bohr's model of the atom failed to address why an atom does not emit radiation when it is in its ground state, what happens when an atom passes from one stationary state to another, or what laws determine the probability of these transitions. In a paper published in 1917 [*Zur Quantentheorie der Strahlung*. (The Quantum Theory of Radiation.)], Albert Einstein addressed the interaction between matter and radiation by means of *absorption* and *emission*, and through incident and outgoing *radiation*.

In December 1921, Rudolf Ladenburg published a paper [*Die quantentheoretische Deutung der Zahl der Dispersionselektronen*. (The quantum-theoretical interpretation of the number of dispersion electrons.)] in which he equated the classical expression for the strength of an absorption line with the quantum-theoretical expression by replacing the atom as far as its interaction with the *radiation field* is concerned by a set of harmonic oscillators with frequencies equal to the absorption frequencies of the atom.

Classical electrodynamics predicts that the energy scattered by an electron traversed by an X-ray beam is independent of the wave-length of the incident rays, it also predicts that when the X-rays traverse a thin layer of matter the intensity of the scattered radiation on the two sides of the layer should be the same. Experiments on the scattering of X-rays by light elements showed that these predictions are correct when X-rays of moderate hardness are employed, but when very hard X-rays or γ-rays are employed, the scattered energy is less than the theoretical value predicted by J. J. Thomson's classical theory of the scattering of X-rays, and is strongly concentrated on the emergent side of the scattering plate.

In a paper published in May 1923 [*A Quantum Theory of the Scattering of X-rays by Light Elements.*], Arthur Compton, at age 31, applied Einstein's hypothesis to the scattering of X-ray and γ-ray photons by electrons and derived the mathematical relationship between the shift in wavelength and the scattering angle of the X-rays *by assuming that each scattered X-ray photon interacts with only one electron*. This agreed with experimental results for the scattering of X-ray and γ-ray photons by electrons, subsequently known as *Compton scattering*, which was important evidence for quantum theory.

In a paper published in May 1924 [*The law of dispersion and Bohr's theory of spectra.*], Hendrick Kramers, at age 29, extended Ladenburg's calculation, by deriving the dispersion (scattering) formula for electromagnetic radiation incident on an atom by also assuming that incident *radiation* was characterized by a train of polarized harmonic waves in which positive virtual oscillators corresponded to *absorption* frequencies, but adding a term representing negative virtual oscillators that corresponded to *emission* frequencies.

In February 1925, Louis de Broglie, at age 32, published a paper [*Recherches sur la théorie des quanta.* (On the Theory of Quanta.)] which described a *relativistic* theory of *wave mechanics* for a moving particle. He applied Einstein's *equivalence of mass and energy* and *relativistic change of mass when moving relative to the observer* to an electron to obtain the *total energy*, and set the *energy* of electron in the rest frame equal to a quantum of energy with a frequency given by Planck's *quantum relationship*. He calculated the *frequency of moving electron* measured by fixed observer by applying *clock retardation*. He resolved the difference from the frequency calculated from the *quantum relation* by showing that the phases of the moving electron and its associated *wave* remained the same. He then represented the wave as a *phase wave with a velocity greater than the velocity of light*, and applied this to the periodic motion of an electron in a Bohr atom and to the mutual interaction of electrons and protons in the hydrogen atom.

In December 1925, Yakov Frenkel, published a paper [*Zur elektrodynamik punktförmiger Elektronen.* (On the electrodynamics of point-like electrons.)] which argued that electrons and protons should be treated as *point charges* and their *masses* considered to be a primary property independent of *charge* in place of *electromagnetic theory of mass*. This was based on erroneous assumption of *mass defect* of helium relative to hydrogen at time when neutron had not been discovered, but he claimed that *an extended electron is inconceivable in the special theory of relativity since due to the intrinsic connection between space and time an invariant definition of a geometrically invariable (i.e., rigid) body is impossible for arbitrary motions*. He noted that electrons are not only physically but geometrically indivisible and have no extension in space, and that there are no internal forces between the elements of an electron because such elements do not exist. The electromagnetic explanation for *mass* then goes away.

The unquestioned acceptance of this led to the unresolved problem of divergencies in the relativistic quantum electrodynamics theory.

Einstein's *theory of special relativity* is based on two postulates:
(i)   The laws of physics are invariant (that is, identical) in all inertial frames of reference (that is, frames of reference with no acceleration), known as the principle of relativity.

(ii)   The speed of light in vacuum is the same for all observers, regardless of the motion of the light source or observer.

It is a theory about observations and laws of physics in inertial frames of reference, that is reference frames moving at constant velocity relative to each other. It is not clear why the

10

radius of an electron (or proton) should be treated as *geometrically invariable* under all frames of reference any more than, say, a foot rule composed of many atoms. Furthermore, neither the position of an electron, nor its radius, are observables under the theory of quantum electrodynamics.

> [Tomonaga, in his 1965 Nobel lecture, noted that "in order to overcome the difficulty of an infinitely large electromagnetic mass, Lorentz considered the electron not to be point-like but to have a finite size. It is very difficult, however, to incorporate a finite sized electron into the framework of relativistic quantum theory".]

In July 1925, at the age of 24, Werner Heisenberg published his seminal paper [*Über quantentheoretische Umdeutung kinematischer und mechanischer Beziehungen.* (On the quantum-theoretical re-interpretation of kinematic and mechanical relations.)] proposing a *new quantum mechanics* in which only relationships among observable quantities occur. It is not possible to assign to the electron a point in space as a function of time, so, building on Kramer's dispersion theory, he assigned to the electron an *emitted radiation*, and substituted *frequencies* and *amplitudes* of the Fourier components of emitted radiation of electron for position in space as a function of time, and assigned *transition frequencies* and *transition amplitudes* as observables. He then translated the old *quantum condition* that fixes the properties of the *states* to a new condition to calculate the amplitude of a *transition* between two states by replacing the differential by a difference. In the quantum case frequencies combine by multiplying *transition amplitudes* (equivalent to matrix multiplication) resulting in the non-commutativity of kinematical quantities. He showed the simple quantum theoretical connection to Kramers' dispersion theory. The *equation of motion* $x'' + f(x) = 0$ and the *quantum condition*
$h = 4\pi m \sum_{\alpha=-\infty}^{+\infty} \{|a(n, n + \alpha)|^2 \omega(n, n + \alpha) - |a(n, n - \alpha)|^2 \omega(n, n - \alpha)\}$ together contain, if solvable, *a complete determination not only of the frequencies and energies but also of the quantum theoretical transition probabilities*.

Max Born realized that Heisenberg's kinematical rule for multiplying position quantities was equivalent to the mathematical rule for multiplying matrices which was unknown to Heisenberg at that time.

In December, 1925, Born and his student Pascual Jordan, at the age of 23, published a paper [*Zur Quantenmechanik.* (On Quantum Mechanics.)] which restated the *commutation relation* in matrix formulation. They derived the *quantum condition* in matrix form
$\mathbf{pq} - \mathbf{qp} = h/2\pi i \, \mathbf{1}$ where $\mathbf{1} = (\delta_{nm})$ with $\delta_{nm} = 1$ for $n = m$; $\delta_{nm} = 0$ for $n \neq m$ and the *equations of motion* $\mathbf{q}\cdot = \delta H/\delta \mathbf{p}$, $\mathbf{p}\cdot = -\delta H/\delta \mathbf{q}$ from the principle of variation. They provided proof that, due to Heisenberg's *quantum condition*, the *energy theorem* and *Bohr's frequency condition* follow from the *equations of motion*. They also showed that the basic laws of the electromagnetic field in a vacuum can readily be incorporated, and provided support for Heisenberg's assumption that the squares of the absolute values of the elements in a matrix

representing the *electrical moment* of an atom provide a measure of the *transition probabilities*.

The idea of a quantized spinning of the electron was put forward for the first time by Compton in August 1921 [*The magnetic electron.*], who pointed out the possible bearing of this idea on the origin of the natural unit of magnetism.

Without being aware of Compton's suggestion Samuel Goudsmit and George Uhlenbeck published a joint paper in November 1925 [*Ersetzung der Hypothese vom unmechanischen Zwang durch eine Forderung bezuglich des inneren Verhaltens jedes einzelnen Elektrons.* (Replacement of the hypothesis of unmechanical coercion by a requirement regarding the internal behavior of each individual electron.)], which noted doublets in the alkali spectra which did not conform to current models of the atom. They proposed applying the model of the spinning electron to interpret a number of features of the quantum theory of the *anomalous Zeeman effect*. They applied the classical formula for a spherical rotating electron with finite radius and surface charge.

In December 1925, Dirac, at the age of 23, published a paper [Dirac, P.A.M. (1925). *The Fundamental Equations of Quantum Mechanics.*] which, following Heisenberg, described the quantization of the electromagnetic field in terms of an ensemble of harmonic components. He assumed that multiplication of quantum variables was not commutative, and called the quantity with components $xy(nm) = \Sigma_k x(nk)y(km)$ the *Heisenberg product* of x and y. He represented this using *Poisson brackets* that occur in the classical dynamics of particle motion, and assumed that the difference between Heisenberg products of two quantum quantities was equal to $ih/2\pi$ times their Poisson bracket expression giving the *quantum condition* $xy - yx = ih/2\pi$. [x, y].

In March 1926, Dirac published a paper [*Quantum Mechanics and a preliminary investigation of the hydrogen atom.*] which applied his *non-relativistic* quantum mechanics to the orbital motion of the electron in the hydrogen atom using Heisenberg's non-communitive quantum variables as *q-numbers*. He used the *quantum conditions* to define *q-numbers*, and *transition frequencies* and *amplitudes* to represent *q-numbers* by means of *c-numbers*. A dynamical system on the classical theory was determined by a Hamiltonian function of p's and q's where the *equation of motion* is expressed in Poisson brackets. Dirac assumed *equations of motion* on the quantum theory of same form where the Hamiltonian is a q number. He stated that a dynamical system on the quantum theory was *multiply periodic* where *uniformizing variables* (action variable J_r and angle variable ϖ) are canonical variables, the Hamiltonian is a function of the J's only, and the original p's and q's are multiply periodic functions of ϖ's of period 2π. He described the Hamiltonian for orbital motion of the electron in the hydrogen atom, and used this to calculate the *transitional frequencies*.

In the same month, Erwin Schrodinger published in German the first of his papers [March, 1926. *Quantisierung als Eigenwertproblem. (Erste Mitteilung).* (Quantization as an eigenvalue problem. (First communication).), subsequently reported in English in a paper published in December 1925 [*A Wave Theory of the Mechanics of Atoms and Molecules.*], addressing his *non-relativistic* development of de Broglie's *relativistic* wave mechanics in which *phase-waves* were associated with the motion of material points, and in particular with the motion of an electron or proton. He assumed that material points are wave-systems with *wave-equation* $\Delta\psi + 8\pi^2 m(E - V)\psi/h^2 = 0$. The *laws of motion* and *quantum conditions* were deduced simultaneously from the Hamiltonian principle. The *wave function* converted an atom into a system of fluctuating *charges* spread out continuously in space, generating an *electric moment* that changed in time. The frequency of emission coincided with differences of the frequency of motion. A definite localization of *electric charge* in space and time was associated with the wave-system. Solutions of the *wave equation* for a simplified hydrogen atom or one body problem corresponded to Bohr's stationary energy levels of the elliptic orbits. The selected values were called *"eigenvalues"* and the solutions that belong to them *"eigenfunctions"*. The charge of the electron was spread out through space but the *wave-phenomenon* was restricted to a small sphere of a few Angstroms diameter constituting the atom. He showed how it was possible to calculate *amplitudes* of harmonic components of the *electric moment* for any direction in space. *Wave mechanics* was developed without reference to *relativity* modifications of classical mechanics or to *action* of a magnetic field on the atom. It had not been possible to extend the *relativistic* theory to a system of more than one electron. He noted that the *relativistic theory of the hydrogen atom was in grave contradiction with experiment.* How to take into account electron spin was yet unknown.

Immediately after Uhlenbeck and Goudsmit published their hypothesis, Heisenberg observed that their explanation of the *anomalous Zeeman effect* based on the spin of the electron produced a precession equal to twice the observed precession.

In April 1926, Llewellyn Thomas published a paper [*The Motion of the Spinning Electron.*] which applied a *relativistic* correction to Uhlenbeck and Goudsmit's hypothesis of electron spin to explain *anomalous Zeeman effect.*

> [It appears highly suspect that applying a Lorentz transformation to the motion of the electron results in halving the rate of precession.]

In May 1926, Dirac published a paper [*The elimination of the nodes in quantum mechanics.*] aimed at simplifying the *non-relativistic* quantum treatment of the dynamical problem of a number of particles or electrons moving in a central field of force and disturbing one another by the introduction of *quantum variables*. In the classical treatment an initial simplification is made, known as the *elimination of the nodes*. This consisted in obtaining a *contact transformation* from the Cartesian co-ordinates and momenta of the electrons to a set of canonical variables of which all except three are independent of the orientation of the system as a whole while these three determined the orientation. The laws of classical mechanics must be generalized when applied to atomic systems; the *commutative law of multiplication* as applied to dynamical variables was replaced by

13

quantum conditions which are just sufficient to enable one to evaluate xy − yx when x and y are given. It followed that the dynamical variables cannot be ordinary numbers expressible in the decimal notation (which numbers will be called *c-numbers*), but may be considered to be numbers of a special kind (which will be called *q-numbers*), whose nature cannot be exactly specified, but which can be used in the algebraic solution of a dynamical problem in a manner closely analogous to the way the corresponding classical variables are used. He introduced *action variables and their canonical conjugate angle variables*, and *transformation equations*, substituted a set of *c-numbers* for *action variables* to fix the *stationary state* and obtain physical results, and applied this to the *anomalous Zeeman effect*. Assuming a ratio of magnetic moment to mechanical angular momentum (g-factor) for the electron is 2, and adopting the usual model of the atom, he showed that *non-relativistic* theory gave the correct g-formula for *energy* of stationary states and Kronig's results for the relative intensities of the lines of a multiplet and their components in a weak magnetic field.

In June of the same year, Dirac published another paper [*Relativity Quantum Mechanics with an Application to Compton Scattering.*] with the object of extending quantum mechanics to systems for which the Hamiltonian involves the time explicitly and to comply with the *theory of special relativity* by treating time on the same footing as the other variables. He set $x_4 = ict$ (so that $x_1^2 + x_2^2 + x_3^2 + x_4^2 = 0$ and $x_1^2 + x_2^2 + x_3^2 = c^2t^2$) and $p_4 = iW/c$ where W is the energy, and showed that −W was the *momentum* conjugate to t. He then substituted $(t − x_1/c)$ for t as a *uniformizing variable* in order that its contribution to the exchange of energy with the radiation field may vanish. He then applied this *relativistic* quantum mechanics to Compton scattering and to the calculation of the *frequency* and *intensity* of scattered radiation. There was no improvement in agreement with experiments from this *relativistic* formulation.

In December 1926, following Einstein, Born published a paper [*Zur Quantenmechanik der Stoßvorgänge.* (On the quantum mechanics of collision processes)] formulating the now-standard interpretation of the *probability density function* for ψ*ψ in the Schrödinger equation, for which he was awarded the Nobel Prize in 1954, "especially for his statistical interpretation of the wavefunction". By studying the collision processes, he developed quantum mechanics in Schrödinger's form that described not only the stationary states, but also the quantum leaps. In his Nobel Prize lecture Born described the background to his 1926 paper. An idea of Einstein's gave him the lead. He had tried to make the duality of particles - light quanta or photons - and waves comprehensible by interpreting the square of the optical wave amplitudes as the *probability density* for the occurrence of photons. This concept could at once be carried over to the ψ-function, $|ψ|^2$ ought to represent the *probability density* for electrons (or other particles). The atomic collision processes suggested themselves at this point. A swarm of electrons coming from infinity represented by an incident wave of known intensity (i.e., $|ψ|^2$) impinges upon an obstacle. The incident electron wave is partially transformed into a secondary spherical wave whose amplitude of oscillation ψ differs for different directions. The square of the amplitude of this wave at a

14

great distance from the scattering center determined the relative probability of scattering as a function of direction. If the scattering atom itself was capable of existing in different *stationary states*, then Schrödinger's wave equation gave automatically the probability of excitation of these states, the electron being scattered with loss of energy - that is to say inelastically. In this way it was possible to get a theoretical basis for the assumptions of Bohr's theory which had been experimentally confirmed by Franck and Hertz. [Franck, J. & Hertz, G. (1914). *Über Zusammenstöße zwischen Elektronen und Molekülen des Quecksilberdampfes und die Ionisierungsspannung desselben.* [On the collisions between electrons and molecules of mercury vapor and the ionization potential of the same.].

In October 1926, Dirac published a paper [*On the Theory of Quantum Mechanics.*] in which he developed a *relativistic* treatment of Schrodinger's wave theory from a more general point of view in which the time t and its conjugate momentum –W were treated from the beginning on the same footing as the other variables. He applied his relativistic formulation to a system containing an atom with two electrons and found that if the positions of the two electrons were interchanged the new state of the atom was physically indistinguishable from the original one. In order that that the theory only enables calculation of *observable quantities* must treat (*mn*) and (*nm*) as only one *state*. *Unsymmetrical* functions of the co-ordinates (and momenta) of the two electrons cannot be represented by matrices. *Symmetrical* functions such as the total *polarizations* of the atom can be considered to be represented by matrices without inconsistency. These matrices are by themselves sufficient to determine all the physical properties of the system. *Theory of uniformizing variables introduced by the author can no longer apply.* New theory allows two solutions satisfying necessary conditions; one leads to Pauli's principle that not more than one electron can be in any given orbit, and the other, when applied to the analogous problem of the ideal gas, leads to the *Einstein-Bose statistical mechanics*. With *neglect of relativity mechanics* accounted for the absorption and stimulated emission of radiation and showed that the elements of the matrices representing the total polarization determined the *transition probabilities. Cannot be applied to spontaneous emission.*

Heisenberg and Schrödinger provided alternative methods for the determination of quantum *frequencies* and *intensities*. The Compton effect was already calculated by Dirac [(June 1926). *Relativity Quantum Mechanics with an Application to Compton Scattering.*] using the Heisenberg method.

In January 1927, Walter Gordon published a paper [*Der Comptoneffekt nach der Schrödingerschen Theorie.* (The Compton effect according to Schrödinger's theory.)] in which the same problem was treated by the Schrödinger method. He started with the same *classic relativistic equation for kinetic energy* in terms of *momentum* and *energy* which is *Hamiltonian equation* for the system $E^2 = p^2c^2 + m^2c^4$, and applied it in same way to *electron in electromagnetic field* described in terms of *vector potential* and *scalar potential*. He then added the same *field energy* to the *kinetic energy* resulting in the same *classical relativistic Hamiltonian equations for a point electron moving in an electromagnetic field.*

In accordance with Schrödinger's rules Gordon then substituted the classical *quantum differential operators* for the momentum vector in the amended *Hamiltonian equation* and applied the resulting differential operator to the *wave function* ψ to obtain the *Klein-Gordon equation*, $1/c^2 \, \partial^2/\partial t^2 \, \psi - \nabla^2 \, \psi + m^2 c^2/h^2 \, \psi = 0$. He then calculated the radiation from the *current density* and *charge density*, and applied this to the Compton effect.

Dirac [(February, 1928). *The Quantum Theory of the Electron.*] objected to this on grounds of the interpretation of the wave function and solutions with negative probabilities and negative energy, and positive charge for the electron.

Heisenberg's original matrix mechanics assumed that the elements of the diagonal matrix that represents the energy are the *energy levels* of the system, and the elements of the matrix that represents the total polarization, which are periodic functions of the time, determine the *frequencies* and *intensities* of the spectral lines in analogy to classical theory. In *Schrodinger's wave representation* physical results are based on assumption that the square of the *amplitude* of the wave function can be interpreted as a probability. This enables the probability of a *transition* being produced in a system by an arbitrary external perturbing force to be worked out.

In January 1927, Dirac published a paper [*The Physical Interpretation of the Quantum Dynamics.*] on *non-relativistic* matrix mechanics providing a *general theory of obtaining physical results from quantum theory*. It showed all the physical information that one can hope to get from quantum dynamics and provided a general method for obtaining it, replacing the special assumptions previously used. This required a theory of the more general schemes of matrix representation in which the rows and columns referred to any set of constants of integration that commute and of the laws of transformation from one such scheme to another. It counted a time variable wherever it occurs as a parameter (a c-number), introduced *transformation equations* that satisfied the *quantum conditions* and *equations of motion*. *Eigenfunctions* of Schrödinger's wave equation were *transformation functions* that enabled transformation from a scheme of matrix representation to a scheme in which Hamiltonian was a diagonal matrix, and dynamical variables were represented by matrices whose rows and columns referred to the initial values of the *action variables* or to the *final values*. The coefficients that enabled transformation from one set of matrices to the other were those that determined the *transition probabilities*.

In October, 1927, Oskar Klein published an alternative calculation [*Elektrodynamik und Wellenmechanik vom Standpunkt des Korrespondenzprinzips.* (Electrodynamics and wave mechanics from the standpoint of the correspondence principle.)] of the Compton effect restricted to the *one-electron problem* starting from the Maxwell-Lorentz field equations. The motion of an electron in an electromagnetic field was described by a *four-potential* and a *scalar potential*. The *Hamilton-Jacobi differential equation* for the *action* function (Klein–Gordon equation) was regarded as the expression for the motion of the electron. Following de Broglie and Schrodinger, Klein replaced this first order equation with a second-order linear equation which represented the *relativistic* generalization of Schrödinger's wave equation for one-electron problem. He then evaluated the equations determining the electromagnetic field with the help of wave mechanics using the correspondence principle to determine wave-mechanical expressions for *electric density* and *current vector*. After *neglecting relativity*, this resulted in the same expressions as those obtained by Schrodinger. Applied to a "bound" electron moving in an axially symmetric electrostatic field over which a weak, homogeneous magnetic field was superimposed, he derived the normal Zeeman effect; and applied to the scattered radiation from a light wave on a "force-free" electron, he obtained the Compton effect.

The new quantum theory, based on the assumption that the dynamical variables do not obey the commutative law of multiplication, had by now been developed sufficiently to form a fairly complete theory of *dynamics*. One could treat mathematically the problem of any dynamical system composed of a number of particles with instantaneous forces acting between them, provided it was describable by a Hamiltonian function, and one could interpret the mathematics physically by a quite definite general method. On the other hand, hardly anything had been done up to the present on *quantum electrodynamics*.

In March 1927, Dirac published a paper [*The quantum theory of the emission and absorption of radiation.*] addressing *non-relativistic quantum electrodynamics*. He noted that the questions of the correct treatment of a system in which the forces are propagated with the velocity of light instead of instantaneously, of the production of an electromagnetic field by a moving electron, and of the reaction of this field on the electron had not yet been touched. In addition, *there was a serious difficulty in making the theory satisfy all the requirements of the restricted principle of relativity* since a Hamiltonian function can no longer be used. It appeared to be possible to build up a fairly satisfactory theory of the *emission of radiation* and of the *reaction of the radiation field on the emitting system* on the basis of a kinematics and dynamics *which were not strictly relativistic*. This was the main object of the paper. *The theory was non-relativistic only on account of the time being counted throughout as a c-number, instead of being treated symmetrically with the space co-ordinates.* The relativity variation of *mass* with *velocity* was taken into account without difficulty.

Dirac treated the problem of an assembly of similar systems satisfying the Einstein-Bose statistical mechanics which interact with another different system by obtaining a Hamiltonian function to describe the motion. The theory was applied to the interaction of an assembly of *light-quanta* with an atom, and it was shown that it led to *Einstein's laws for the emission and absorption of radiation*. The interaction of an atom with *electromagnetic waves* was then considered. The *field* of radiation was treated as a dynamical system whose interaction with an ordinary atomic system might be described by a Hamilton function. The dynamical variables specifying the *field* were the *energies* and *phases* of the harmonic components of the waves; *q-numbers* satisfying the proper quantum conditions instead of *c-numbers*. Then the Hamiltonian function took the same form as that

17

for the interaction of an assembly of light quanta with the atom. This provided a complete formal reconciliation between the wave and light-quantum point of view. The theory *led to the correct expressions for Einstein's A's and B's*. The mathematical development of this theory was made possible by Dirac's *general transformation theory* of the quantum matrices [Dirac (January, 1927). *The Physical Interpretation of the Quantum Dynamics*].

In May 1927, Dirac published a paper [*The quantum theory of dispersion.*] in which he applied his *non-relativistic quantum electrodynamics* theory in the previous paper [Dirac (March, 1927). *The quantum theory of the emission and absorption of radiation.*] to determine the radiation scattered by the atom. The method used involved finding a solution of Schrodinger equation that satisfied initial conditions corresponding to a given *initial state* for the atom and field. Scattered radiation appeared as result of two processes, an a*bsorption* and an *emission*. The problem of light quanta being emitted not converging at high frequencies arose from using an approximation of regarding the atom as a dipole, but use of the exact expression for *interaction energy* was too complicated for radiation theory at that time. *This led to the correct formula for scattering of radiation by a free electron, with neglect of relativity*, and thus of the Compton effect. The approximation was sufficient for *dispersion* and *resonance* but inadequate to calculate the *breadth of a spectral line*.

In September 1927, Wolfgang Pauli published a paper [*Zur Quantenmechanik des magnetischen Elektrons.* (On the quantum mechanics of magnetic electrons.)] that showed how the *non-relativistic* formulation by Dirac [Dirac (January, 1927). *The Physical Interpretation of the Quantum Dynamics*] and Jordan using the general canonical transformations of the Schrödinger functions enabled a quantum-mechanical representation of electrons by the method of *eigenfunctions*. The differential equations for the *eigenfunctions* of the magnetic electron given in this paper were only provisional and approximate since they, like the Heisenberg-Jordan matrix formulation, *were not written down in a relativistically-invariant way*. For the hydrogen atom they were valid only in the approximation in which the dynamical behavior of the proper moment could be considered to be a secular perturbation.

In a paper published in February 1928 [*The Quantum Theory of the Electron.*], Dirac noted that the new quantum mechanics applied to the problem of the structure of the atom with *point-charge electrons* did not give results in agreement with experiment. The discrepancies consisted of "duplexity" phenomena; the observed number of stationary states for an electron in an atom being twice the number given by the theory. Goudsmit and Uhlenbeck [Goudsmit, S. & Uhlenbeck, G. (1925). *Ersetzung der Hypothese vom unmechanischen Zwang durch eine Forderung bezuglich des inneren Verhaltens jedes einzelnen Elektrons.* (Replacement of the hypothesis of unmechanical coercion by a requirement regarding the internal behavior of each individual electron.)] introduced the idea of an electron with a *spin*. Previous *relativity* treatments by Gordon and Klein

[Gordon, W. (1927). *Der Comptoneffekt nach der Schrödingerschen Theorie.* (The Compton effect according to Schrödinger's theory.) and Klein, O. (1927). *Elektrodynamik und Wellenmechanik vom Standpunkt des Korrespondenzprinzips.* (Electrodynamics and wave mechanics from the standpoint of the correspondence principle.)] obtained the operator of the wave equation by the same procedure as in the *non-relativity* theory; they substituted classical *quantum differential operators* for the *momentum vector* in the amended *relativistic Hamiltonian equation* and applied the resulting differential operator to the *wave function* to obtain the *Klein-Gordon equation.* Dirac noted that Gordon and Klein's treatments gave rise to two difficulties. The *first difficulty* was in the physical interpretation of solutions of ψ as the *charge* and the *current*. This was satisfactory for emission and absorption of radiation, but only provided the probability of any dynamical variable at any specific time having a value between specified limits if they referred to the *position* of the electron, but, unlike the *non-relativity* theory, *not if they refer to its momentum or any other dynamical variable.* The *second difficulty* was that the conjugate imaginary of the wave equation was the same as that for an electron with charge – e and negative energy.

This paper only addressed the removal of the first of difficulties. The resulting theory was only an approximation but appeared sufficient to address the duplexity problems without further assumptions. Dirac applied the method of *q-numbers* and using non-commutative algebra exhibited the properties of a free electron and of an electron in a central field of electric force. He showed that simplest Hamiltonian for a *point charge electron satisfying requirements of both relativity and the general transformation theory* of quantum mechanics led to an explanation of all duplexity phenomena of number of stationary states being twice the observed value *without further assumption about spin.* In contrast to the Schrödinger equation which described wave functions of only one complex value, Dirac introduced *vectors of four complex numbers* (known as bispinors). This resulted in a *relativistic equation of motion* for the *wave function of the electron* referred to as the *Dirac equation,* $\{p_0 + \rho_1 (\boldsymbol{\sigma}, \mathbf{p}) + \rho_3 mc\} \, \psi = 0$, where $\mathbf{p}$ is the *momentum* vector, and $\boldsymbol{\sigma}$ denotes the vector $(\sigma_1, \sigma_2, \sigma_3)$. This included a term equal to the spin correction given by Darwin and Pauli. It described all spin-½ particles with mass, but did not address the second class of solutions of the wave equation in which *charge of the electron is positive* and *energy of a free electron is negative.*

In the second part of this paper, published in March 1928 [*The Quantum Theory of the Electron. Part II.*], Dirac applied the *Dirac equation* to the conservation theorem, the selection principle, the relative intensities of the lines of a multiplet, and to the Zeeman effect.

In a paper published in April 1929 [*Quantum Mechanics of Many-Electron Systems.*], Dirac noted that the general theory of quantum mechanics was now almost complete; *the imperfections that still remained being in connection with the exact fitting in of the theory with relativity ideas.* These gave rise to difficulties *only when high-speed particles were involved* and were therefore of no importance in the consideration of atomic and molecular structure and ordinary chemical reactions. *The difficulty was only that the exact application of these laws led to equations much too complicated to be soluble.* He considered it to be desirable that approximate practical methods of applying quantum mechanics should be developed which could lead to an explanation of the main features of complex atomic systems without too much computation.

With the help of the spin of the electron and Pauli's exclusion principle, a satisfactory theory of multiplet terms was obtained when the additional assumption was made that the electrons in an atom all set themselves with their spins parallel or antiparallel, but there was no theoretical reason to support this. This seemed to show that there were large forces coupling the *spin vectors* of the electrons in an atom. Dirac provided an explanation based on the *exchange interaction* of electrons arising from electrons being indistinguishable one from another, which resulted in large *exchange energies* between electrons in different atoms and accounted for homopolar valency bonds. For each *stationary state* of the atom there was one magnitude of the total spin vector. He noted that developments of the *theory of exchange* made by Heitler, London and Heisenberg made extensive use of *group theory,* a theory of certain quantities that do not satisfy the commutative law of multiplication, which should thus form a part of quantum mechanics. Dirac translated the methods and results of *group theory* into the language of *quantum mechanics*. He showed that the *exchange interaction* was equal to a constant *perturbation energy* together with *coupling energy* between spin vectors, which determined energy levels. He also showed that, in the first approximation, the *exchange interaction* between the electrons could be replaced by a coupling between their spins; and that the energy of this coupling for each pair of electrons was equal to the scalar product of their *spin vectors* multiplied by a numerical coefficient given by the *exchange energy*.

In the years 1926–1928, immediately following the creation of matrix and wave mechanics, the protagonists of this development elaborated and expanded the techniques of the new quantum mechanics, to apply them to *field theories*.

This work culminated in the publication in July 1929 by Heisenberg and Pauli of their first attempt [Heisenberg, W. & Pauli, W. (1929). *Zur Quantendynamik der Wellenfelder.* (On the quantum dynamics of wave fields.)] to construct their own version of a *relativistically invariant quantum electrodynamics* to treat the interaction between matter and the electromagnetic *field* and between matter and matter. This paper dealt mainly with the *canonical quantization of both the electromagnetic and the matter-wave fields*, but this theory led to *divergent expressions for the energies of stationary states* and *the differences between these energies* (i.e., the actually observed frequencies of spectral lines) *came out*

20

infinite. In order to write down a Lorentz-invariant Lagrangian for the interacting *electromagnetic* and *matter-wave fields* it was necessary to work with the *electromagnetic potentials* and not just with the *fields*. But the Lagrangian does not contain a time derivative of the *electric potential* so that there is no corresponding canonical momentum variable, preventing the straightforward implementation of canonical commutation relations. *The theory was also not manifestly covariant* due to the use of equal-time commutation relations. *The fundamental difficulties in the relativistic formulation that were emphasized by Dirac remained* unchanged. The formulas of the theory led to an *infinite zero-point energy* for the radiation and thus included the interaction of an electron with itself as an *infinite* additive constant. However, these difficulties were of a sort that they did not interfere with the application of the theory to many physical problems. This paper used a "crude trick" of adding additional terms to the Lagrangian.

A follow-up paper published in January 1930 [[Heisenberg, W. & Pauli, W. (1930). *Zur Quantendynamik der Wellenfelder II.* (On the quantum dynamics of wave fields II.)], applied a new approach to Lorentz-invariant Lagrangian problem based on the notion of *gauge invariance* of the *theory of coupled electromagnetic potentials* and Dirac *matter waves*. Integrals of the *equations of motion* were derived from *invariance properties* of Hamiltonian function, and the *invariance properties* of *wave equations* were exploited in a similar way. An *infinite* interaction of the electron with itself also resulted from this approach making application of the theory impossible in many cases. The theory led to divergent expressions for the *energies* of *stationary states* and the differences between these *energies* (i.e., the actually observed frequencies of spectral lines) came out *infinite*.

In September 1931, Dirac published a paper [*Quantized singularities in the electromagnetic field.*] of which the object was to show that quantum mechanics did not preclude the existence of *isolated magnetic poles*. He addressed the reason for the *smallest electric charge* e known experimentally to be given by $hc/e^2 = 137$. Using *non-relativistic* theory, he considered a particle whose motion was represented by a wave function and showed that the *change in phase* round a closed curve must be the same for all *wave functions*. He then applied this to the motion of an electron in an electromagnetic field to show that the non-integrable derivatives of the phase of the wave function represented *potentials* of the electromagnetic field. This connection between *non-integrability of phase* and *electromagnetic field* was essentially Weyl's *principle of gauge invariance*. This led to wave equations whose only physical interpretation was the motion of an electron in the field of a single pole. It did not give a value for e but showed reciprocity between *electricity* and *magnetism*; and the strength of a pole and the electric charge must both be quantized. It also gave the relationship between the strength of quantum of a magnetic pole and the electronic charge $hc/e\mu_0 = 2$ but *did not explain their magnitudes*. The reason that isolated magnetic poles had not been separated was probably due to the very large force between two one-quantum poles of opposite sign, $(137/2)^2$ times that of that between an electron and a proton.

21

Volume II.

In 1925, when he was 24 years old, Heisenberg formulated a type of quantum mechanics based on matrices [Heisenberg, W. (July, 1925). *Über quantentheoretische Umdeutung kinematischer und mechanischer Beziehungen.* (On the quantum-theoretical re-interpretation of kinematic and mechanical relations.)]. In 1927 he proposed the "uncertainty relation", setting limits for how precisely the position and velocity of a particle can be simultaneously determined. On December 11, 1933, Werner Heisenberg received the 1932 Nobel Prize in Physics "for the creation of quantum mechanics, the application of which has, inter alia, led to the discovery of the allotropic forms of hydrogen". In his Nobel Prize lecture [Werner Heisenberg – 1932 Nobel Lecture, December 11, 1933. *The development of quantum mechanics.*], he noted that "the impossibility of harmonizing the Maxwellian theory with the pronouncedly visual concepts expressed in the hypothesis of light quanta subsequently compelled research workers to the conclusion that *radiation phenomena can only be understood by largely renouncing their immediate visualization.* … Classical physics seemed the limiting case of visualization of a fundamentally unvisualizable microphysics, the more accurately realizable the more Planck's constant vanishes relative to the parameters of the system". After describing the development of the current theory of quantum mechanics, he concluded that "a visual description for the atomic events is possible only within certain limits of accuracy - but within these limits the laws of classical physics also still apply. Owing to these limits of accuracy as defined by the uncertainty relations, moreover, a visual picture of the atom free from ambiguity has not been determined. On the contrary the corpuscular and the wave concepts are equally serviceable as a basis for visual interpretation. The laws of quantum mechanics are basically statistical". He noted that "the attention of the research workers was now primarily directed to *the problem of reconciling the claims of the special relativity theory with those of the quantum theory.* … The attempts made hitherto to achieve a *relativistic* formulation of the quantum theory are all based on visual concepts so close to those of classical physics that it seems impossible to determine the fine-structure constant within this system of concepts".

Assuming that matter (e.g., electrons) could be regarded as both particles and waves, in 1926, when he was 39 years old, Schrödinger formulated a wave equation that accurately calculated the energy levels of electrons in atoms [Schrödinger, E. (1926). *A Wave Theory of the Mechanics of Atoms and Molecules.*]. In Niels Bohr's theory of the atom, electrons absorb and emit radiation of fixed wavelengths when jumping between fixed orbits around a nucleus. The theory provided a good description of the spectrum created by the hydrogen atom, but needed to be developed to suit more complicated atoms and molecules. On December 12, 1933, Erwin Schrödinger received the 1933 Nobel Prize in Physics, together with Paul Dirac, "for the discovery of new productive forms of atomic theory". In his Nobel Prize lecture, Schrödinger described a fascinating analogy between the path of a ray of light and the path of a mass point. In optics the old system of mechanics corresponded to operating with isolated mutually independent light rays. The new wave mechanics corresponds to the wave theory of light. We are never in a position to say what really is or what really happens, but we can only say what will be observed in any concrete individual case. He concluded that *the ray or the particle path corresponded to a longitudinal relationship of the propagation process (i.e. in the direction of propagation), the wave surface on the other hand to a transversal relationship (i.e. normal to it). Both relationships were without doubt real;* one is proved by photographed particle paths, the other by interference experiments. *To combine both in a uniform system had proven impossible so far. Only in extreme cases did either the transversal, shell-shaped or the radial longitudinal relationship predominate to such an extent that we think we can make do with the wave theory alone or with the particle theory alone.*

During the intense period of 1925-26 quantum theories were proposed that accurately described the energy levels of electrons in atoms. These equations needed to be adapted to Einstein's theory of relativity, however. In 1928 Dirac, when he was 26 years old, formulated a fully *relativistic* quantum theory [Dirac, P. A. M. (February, 1928). *The Quantum Theory of the Electron.*]. The equation gave solutions that he interpreted as being caused by a particle equivalent to the electron, but with a positive charge. This particle, the positron, was later confirmed through experiments. In his Nobel Lecture, Dirac described the current state of his theory of electrons and positrons. He focused on electrons and the positrons on the grounds that the theory has been developed further, and hardly anything could be inferred theoretically about the properties of the others. He noted that the question that must first be considered is *how theory can give any information at all about the properties of elementary particles.* There existed at the present time a general quantum mechanics which could be used to describe the motion of any kind of particle, no matter what its properties were. The general quantum mechanics, however, was valid only when the particles had small velocities and failed for velocities comparable with the velocity of light, when the effects of *relativity* come in. There existed no *relativistic* quantum mechanics for particles with large velocities which could be applied to particles with arbitrary properties. But, *subjecting quantum mechanics to relativistic requirements, imposed restrictions on the properties of the particle, and in this way information about the particles could be deduced from purely theoretical considerations.* Dirac described how the spin properties of the electron could be deduced, and the existence of positrons with similar spin properties and with the possibility of being annihilated in collisions with electrons could be inferred. The new variables which he introduced to get a *relativistic* wave equation linear in *kinetic energy* W gave rise to the spin of the electron. These variables also gave rise to some unexpected phenomena concerning the motion of the electron. *It was found that an electron which appears to be moving slowly, must actually have a very high frequency oscillatory motion of small amplitude superposed on the regular motion, which resulted in the velocity of the electron at any time being equal to the velocity of light.* The wave equation also allowed negative-energy states in an electromagnetic field which *corresponded to the motion of an electron with a positive charge instead of the usual negative one*, i.e. a positron, which Dirac identified with a "hole" in his "hole theory". On this view, the positron was just a mirror-image of the electron, having exactly the same mass and opposite charge.

In January 1930, Dirac published a paper [*A theory of electrons and protons.*], on which his 1933 Nobel Lecture was based, in which he formulated a fully *relativistic* quantum theory. He noted that in *relativistic* quantum theory, in which the *electromagnetic field* is subjected to quantum laws, the *wave equation* refers equally well to an electron with charge + e with *negative kinetic energy*, a difficulty not restricted to the quantum theory of the electron but one which appears in all relativity theories. It arises because there is an ambiguity in the sign of W, or rather $W + eA_0$, in the *relativity* Hamiltonian equation of the classical theory. In the quantum theory transitions can take place in which the *energy* of the electron changes from a positive to a negative value. Dirac used his "*hole theory*" to address the problem by assuming that in a vacuum all negative-energy electron eigenstates were occupied. He noted that if negative-energy *eigenstates* are incompletely filled each unoccupied *eigenstate* – called a "hole" – would behave like a positively charged particle, which he initially thought might be a proton. He noted that for the scattering of radiation by an electron the *exclusion principle* forbids the electron to jump into a state of *negative energy*, so he assumed a double transition process in which first one of the negative-energy electrons jumps up into the *final state* for the electron with the *absorption* (or *emission*) of a photon, and then the original positive-energy electron drops into hole formed by first transition with *emission* (or *absorption*) of a photon.

In November 1930, Heisenberg published a paper [*Die Selbstenergie des Elektrons.* (The self-energy of the electron.)] in which he noted that a *point-like* electron results in infinite energy density, but a *finite radius* for the electron is inconsistent with Special Relativity. *He chose to represent an electron by a point charge*, and investigated the conditions under which the self-energy of the electron vanished. He showed that the one-electron problem could be treated correctly without an infinite *self-energy* if there were solutions of vacuum electrodynamics without a *zero-point energy*, but *such solutions did not exist*. He showed that the difficulties of *field theory* did not come directly from the infinite *self-energy* of electron, and concluded that *a solution of the basic equations had therefore not been found for the time being,* and that it was also not probable that one would achieve a solution without substantial modification of the quantum theory of wave fields.

In May 1932, Dirac published a paper [*Relativistic Quantum Mechanics.*] describing an alternative to Heisenberg and Pauli's approach to *relativistic* quantum mechanics which regarded the *field* itself as a dynamical system amenable to Hamiltonian treatment and its interaction with the particles as describable by an *interaction energy*. He noted that there were serious objections to these views. *If we wish to make an observation on a system of interacting particles the only effective method of procedure is to subject them to a field of electromagnetic radiation and see how they react.* The role of the *field* is to provide a means for making observations. The nature of an observation requires an interplay between the *field* and the *particles*. We cannot suppose the *field* to be a dynamical system on the same footing as the *particles* and thus something to be observed in the same way as the *particles*. The *field* should appear in the theory as something more elementary and fundamental. In this paper he proposed an *interaction representation* which gave interplay between particles and the field, and he translated the *equations of motion* of *relativistic* classical theory directly into equations expressible entirely in terms of *probability amplitudes* referring to one *ingoing field* and one *outgoing field*. He assumed that passage from the field of ingoing waves to the field of outgoing waves was a quantum jump performed by one field composed of waves passing undisturbed through the electron and satisfying Maxwell's equations. The *relativistic* observable quantities were *transition probabilities*, with *probability amplitudes* analogous to Heisenberg's matrix elements. For quantization he assumed that *intensities* and *phases* were operators satisfying the usual quantum conditions governing the *intensities* and *phases* of Fourier components of electromagnetic field in empty space, which determined the matrix elements associated with electron jumps. He assumed that the interaction of each electron with the *field* could be described by an *interaction energy* equal to its charge multiplied by the *potential* at the point where it was situated, and derived the *wave equation* for interactions of two electrons due to motions of each being connected with same field to be $\{ih\, \delta/\delta t + h^2/2m_1\, \delta^2/\delta x_1{}^2 + h^2/2m_2\, \delta^2/\delta x_2{}^2 - \varepsilon_1 V(x_1 t) - \varepsilon_2 V(x_2 t)\}\, \psi = 0$.

In late 1932, Dirac, Fock, and Podolsky published a paper [*On quantum electrodynamics.*] in the *Phys. Zeit. Sowjetunion* which described a *relativistic* model in which a fixed number of electrons interacted through a second-quantized electromagnetic field. They applied Dirac's *interaction representation* formulation of quantum field theory to the full electrodynamics. This was formulated with the help of a *multi-time* wave function $\psi(t_1,\mathbf{x}_1,\ldots,t_N,\mathbf{x}_N)\psi(t_1,x_1,\ldots,t_N,x_N)$ that generalized the Schrödinger's multiparticle wave function to allow for a manifestly *relativistic* formulation of wave mechanics. All non-trivial dynamics due to interaction between charged matter and electromagnetic field were relegated to the time evolution of a state vector. The second Maxwell equation - the Ampère-Maxwell law – was satisfied only through the action of the field operators on the wave function. This implied that electromagnetic field operators still obeyed the free field equations and consequently the covariant commutation relations of Jordan and Pauli, but the method of quantizing by imposing covariant commutation relations was still an isolated technique applied only to Maxwell field. This showed relationship between *Maxwell's equations* for empty space, the *wave equation* of Heisenberg-Pauli's theory, and the *wave equation of* Dirac's *interaction representation*.

In 1933, Dirac published a paper [*The Lagrangian in Quantum Mechanics.*], in the same journal, which provided an alternative formulation of quantum mechanics in terms of Lagrangian in place of Hamiltonian. This used *coordinates* and *velocities* instead of *coordinates* and *momenta*, and allowed the *equations of motion* to be expressed as stationary property of the *action function*, the time-integral of the Lagrangian, which was

25

a *relativistic invariant*. There is no corresponding *action principle* in terms of *coordinates and momenta* in Hamiltonian theory. Lagrangian theory is closely connected with theory of *contact transformations*. Dirac derived *transformation functions* that had classical analogues expressible in terms of the Lagrangian. The classical Lagrangian is a function of coordinates at time t and time t + dt rather than of the coordinates and velocities. He introduced his *"many-time"* theory and applied it to *field dynamics* using suitable field quantities or potentials as coordinates, in which each coordinate was a function of four *space-time* variables instead of one time variable as in particle theory. This resulted in a quantum analogue of the classical *transformation function* between dynamical variables that Dirac described as a sort of *generalized transformation function*.

In March 1934, Dirac published a paper [*Discussion of the infinite distribution of electrons in the theory of the positron.*] which attempted to address the problem with his *relativistic* 'hole' theory that implied an infinite number of negative-energy electrons (per unit volume) with energies extending continuously from $-mc^2$ to $-\infty$. When an electromagnetic field was present positive- and negative-energy states could not be distinguished in a *relativistically* invariant way. It is necessary to set up assumptions for production of *electromagnetic field* by the electron distribution such that any finite change in distribution produced a change in the field in agreement with Maxwell's equations and such that the infinite field which would be required by Maxwell's equations from an infinite density of electrons is in some way cut out. The exact treatment is very complicated; the present paper gave an approximate treatment. Dirac assumed that *each electron had its own individual wave function in space-time* and *each electron moved in an electromagnetic field which was the same for all electrons*, part coming from external causes and part from the electron distribution itself. He defined the distribution of electrons in terms of the *relativistic density matrix* $\sum_{oc} \psi_{k'}(x' \, t') \, \psi^*_{k''}(x'' \, t'')$ referring to two points in space and two times, and separated the distribution into two parts, where one contained the singularities, and the other described the *electric* and *current densities* physically present.

In March 1934, Heisenberg published a paper [*Bemerkungen zur Diracschen Theorie des Positrons.* (Remarks on the Dirac theory of positron.)] with the purpose of reconstructing Dirac's *theory of the positron* in the formalism of quantum electrodynamics. It demanded that the symmetry in nature between positive and negative charge should be expressed in the basic equations from the outset. In addition to the well-known difficulties with the divergences, no new infinities should appear in the formalism, moreover the theory should provide an approximation for the treatment of the circle of problems that have been treated by quantum electrodynamics up to now. Dirac [(1934). *Discussion of the infinite distribution of electrons in the theory of the positron.*] showed that a quantum mechanical system of many electrons that fulfilled the Pauli principle and moved in a given force field without back-reaction could be characterized by a *density matrix*
$(x', t', k' \mid R \mid x'', t'', k'') = \sum_n \psi^*_n (x', t', k') \, \psi_n (x'', t'', k'')$ where $\psi_n(x', t', k')$ meant the normalized eigenfunctions of the states that possessed one electron, and x', t', k' (x'', t'', k'', resp.) were the position, time, and spin variables. All physically-important properties of

quantum-mechanical systems like *charge density*, *current density*, etc., could be read off from the *density matrix*. The temporal change in the *density matrix* was determined by the Dirac differential equation. Dirac made different choice of *density matrix* representing the external field, resulting in a different energy and impulse density. By restricting oneself to an *intuitive analogue theory of matter fields, the negative energy levels in the Dirac theory could be avoided by replacing the homogeneous Dirac differential equation with an inhomogeneous equation*, where the inhomogeneity was indicative of pair creation. For the most practical applications e.g., pair creation, annihilation, Compton scattering, etc. *this theory described did not yield anything new compared to the formulation of the Dirac theory.* In the Maxwell theory, a continuous charge distribution also led to a finite self-energy; it is the "quantization" that leads to the infinite self-energy. *If one represents the quantization of the electromagnetic field by point-like light quanta then the infinitude of the self-energy also emerges in the intuitive theory of matter waves.*

In January 1934, Victor Weisskopf published a paper [*Über die Selbstenergie des Elektrons.* (The Self-energy of the Electron.)] in which the *self-energy* of the electron was derived in close formal connection to classical radiation theory without direct application of quantum electrodynamics. The radiation field was calculated classically from the *current and charge densities* of the atom. The electromagnetic field was divided into a rotation-free *electrostatic* part and a divergence free *electrodynamic* part. The *self-energy* derived from the electrostatic part. A correction to the paper showed that the *self-energy* of electron at occupied negative energy states diverged logarithmically as in Dirac's "hole theory", in contrast to the linear divergence of the classical theory and the quadratic divergence of the one particle Dirac theory.

The *scalar relativistic wave equation* $E^2/c^2 - \sum_{k=1}^{3} p_k^2 - m^2c^2 = 0$ had generally been relinquished in favor of *Dirac's four-component wave equation* $\{p_0 + \rho_1 (\boldsymbol{\sigma}, \mathbf{p}) + \rho_3 mc\} \psi = 0$ [Dirac, P. A. M. (February, 1928). *The Quantum Theory of the Electron.*] because the former did not yield the spin of the particles.

In July 1934, Pauli and Weisskopf sent a paper for publication [*Über die Quantisierung der skalaren relativistischen Wellengleichung.* (The Quantization of the Scalar Relativistic Wave Equation.)] which argued that Dirac's *a priori* arguments based on limitation to a *single-body problem* and *particle density* being a meaningful observable needed to be revised. It showed that *charge density* is a meaningful observable, that there was no reason for special form of *charge density* $\sum_r \psi^*_r \psi_r$, in the *scalar relativistic wave equation theory*, and that the energy of material particles is always positive after wave fields have been quantized. No new hypothesis such as the "hole theory" was required. It applied Heisenberg-Pauli formalism of quantization of wave fields to the *scalar relativistic wave equation* for matter fields for particles without spin and with Bose-Einstein statistics, and showed that quantization of the *Klein-Gordon relativistic wave equation* for *scalar particles* gave rise to particles with opposite charge but with same rest mass which can be created or destroyed with absorption or emission of electromagnetic radiation. The frequency of pair creation and annihilation processes was of same order of magnitude as the frequency for particles of same charge and mass which followed from Dirac's "hole theory". It *also led to infinite self-energy not only of the particles but also to an infinite polarizability of the vacuum.*

In September 1934, Ernst Stueckelberg, published a paper [*Relativistisch invariante Störungstheorie des Diracschen Elektrons I. Teil: Streustrahlung und Bremsstrahlung.* (Relativistically invariant perturbation theory of Dirac's electron Part I: scattered radiation and Bremsstrahlung.)] in which the main innovation was the introduction of a new perturbative scheme yielding relativistic expressions for the *matrix elements* which were manifestly *gauge invariant*. This was achieved by performing a four-dimensional Fourier transformation of the *wave-function* eliminating space and time variables based on the interaction picture of Dirac, Fock and Podolsky. The starting point was the Dirac equation for the *spinor wave-function* [1/i (γ, $\partial/\partial x$) + M + eV(x)] Ψ(x) = 0. The contributions of positive and negative energies (corresponding to virtual electrons and positrons) were contained in a single propagation function which corresponded to what later was called the *Feynman propagator*. His method for calculating the cross-section started with the definition of *on mass-shell wave-functions* and used integration over the *complex energy-plane*. This paper introduced *Stueckelberg diagrams* later adopted by Richard Feynman and subsequently renamed *Feynman diagrams*. All of Stueckelberg's expressions for matrix elements were identical to those obtained nowadays from Feynman diagrams.

In August 1938, Dirac published a paper [*Classical Theory of Radiating Electrons.*] describing a *relativistic* form of classical theory of radiating electrons assuming the electron to be a *point charge* with no volume. He noted that an extended electron was inconceivable in the *special theory of relativity* due to intrinsic connection between space and time, and that the Lorentz *model of the electron as possessing mass on account of the electromagnetic field around it failed* without further assumptions if the electromagnetic field varied too rapidly or if the acceleration of the electrons was too great. No natural way of introducing further assumptions had been discovered. The discovery of the neutron introduced a form of mass for which it was difficult to believe that it was of electromagnetic nature. Also, the theory of the positron in which positive and negative values of an electron play symmetrical roles could not be fitted into the electromagnetic idea of mass, which required all mass to be positive. The *departure from the electromagnetic theory of mass removed the main reason for believing in a finite size of the electron*. This resulted in the difficulty that the *field* in the immediate neighborhood of the electron had an infinite mass. In quantum mechanics, this resulted in a *divergence* in the solution of the equations that described the interaction of an electron with an electromagnetic field and prevented its application to high-energy radiation processes.

In this paper Dirac chose to represent electron by point charge and to avoid the difficulties with the infinite energy by direct omission or subtraction of unwanted terms, as had been used in the theory of the positron. *The new theory had to be in agreement with special*

28

relativity and the conservation of energy and momentum. Dirac first addressed the problem of a *single electron moving in an electromagnetic field.* He used the equations for the electromagnetic *potentials* in terms of the *charge-current density* vector, and a*ssumed that the charge-current density vector vanished everywhere except on the world-line of the electron, where it was infinitely great.* He derived the *field* quantities from the electromagnetic *potentials,* and obtained solutions in terms of *retarded* and *advanced potentials* and *potentials* representing the incoming and outgoing radiation, from which he calculated the field of radiation produced by the electron. These were in agreement with classical theory but provided a value for the field of radiation throughout space-time. He obtained exact *equations of motion* of the electron in a specified incident *field* from the laws of conservation of energy and momentum within the limits of classical theory, and imposed the condition that solutions which occur in Nature when there is no incident field were those for which the velocity in terms of proper time is constant. This required solutions of the equations of motion *for which the final acceleration as well as the initial position and velocity were prescribed.* The electron responded to a pulse of electromagnetic radiation before it reaches the center of the electron; behaving as if it had a radius of order $1/a$ where $a = 3m/e^2$. This implied that a signal could be transmitted *faster than the speed of light* through the interior of an electron, and that the *interior of electron was a region of failure of some of the elementary properties of space-time.* He showed how the theory could be extended to any number of electrons interacting with each other and with a field of radiation.

In July 1939, Dirac published a paper [*A new notation for quantum mechanics.*] introducing the *bra ket* notation. He noted that quantum mechanics dealt with *vectors in Hilbert space,* representing the *states* of a dynamical system, and with *linear operators,* representing dynamical *variables,* and that sometimes one made calculations using the *vectors and linear operators* directly, treating them as *abstract quantities,* at other times one worked with *coordinates* (or *representatives*) of these *quantities.* The *bra ket* notation provided a concise way of writing the *abstract quantities* themselves and their *coordinates* in a single scheme, which led to a unification of ideas.

In July 1939, Weisskopf published a paper [Weisskopf, V.F. (1939). *On the Self-energy and Electromagnetic Field of the Electron.*] of which t*he main purpose was to show the physical significance of the logarithmic divergence of the self-energy of the electron and to demonstrate the reason for its occurrence.* The self-energy of the electron is its total energy in free space when isolated from other particles or light quanta,
$W = T + (1/8\pi) \int (H^2 + E^2)$ dr where T is the *kinetic energy* of the electron and H and E are the *magnetic* and *electric field strengths* at point r. He identified three reasons why the quantum theory of the electron resulted in the infinite *self-energy* of the electron. *Quantum kinematics required that the radius of the electron must be assumed to be zero,* resulting in infinite energy of the *electrostatic field.* The contributions of the electric and magnetic fields of the spin to the *self-energy* of the electron canceled one another. *Quantum theory of the electromagnetic field postulated the existence of field strength fluctuations in empty*

space, which gave rise to an additional energy that diverged more strongly that the electrostatic self-energy It induced the electron to perform vibrations with energy that diverged quadratically *for an infinitely small radius*. Dirac's positron theory implied that the charge and magnetic dipole of the electron were extended over a finite region. Weisskopf explained why the *self-energy* was only *logarithmically infinite. Divergences were a consequence of the assumption of a point electron.*

On June 19, 1941, Dirac delivered the Bakerian Lecture [*The physical interpretation of quantum mechanics*, published March 1942] at Burlington House, whilst London was being bombed. He described how a satisfactory *non-relativistic* quantum mechanics had been established, in which the *Heisenberg method* focused on quantities which enter into experimental results, and connected together in one calculation probability coefficients from all initial states to all final states; and in which the *Schrodinger method* connected together in one calculation probability coefficients for transitions from one particular initial state to any final state. Both methods rested on same mathematical formalism, a generalization of the Hamilton form of classical dynamics, involving linear operators instead of ordinary algebraic variables. The methods of physical interpretation differed, and was probably still not finally settled. *The theory was not in agreement with the theory of special relativity*, as was evident by the special role played by the time t. While it worked very well in the *non-relativistic* region of low velocities, where it appeared to be in complete agreement with experiment, it could be considered only as an approximation. Setting up the mathematical formalism was fairly straightforward. Firstly, the classical mechanics needed to be put into *relativistic* Hamiltonian form, taking into account that that the various particles comprising the dynamical system interact through the medium of the electromagnetic field, using Lorentz's *equations of motion*, including the dampening terms which expressed the reaction of radiation. This Hamitonian formulation could then be made into a quantum theory following the procedure from *non-relativistic* quantum mechanics. Although this appeared satisfactory mathematically, it met serious difficulties in its physical interpretation. In particular, *it resulted in states of negative energy and negative probability*. It appeared that, whether one was dealing with particles of integral spin or of half-odd integral spin, the mathematical methods at present in use in quantum mechanics *were capable of direct interpretation only in terms of a hypothetical world differing very markedly from the actual one*. All that *relativistic* theory did was provide a consistent means of calculating experimental results.

In July 1943, S-matrix theory was proposed as a principle of particle interactions by Heisenberg [Heisenberg, W. (1943). *Die beobachtbaren Größen in der Theorie der Elementarteilchen. I.* (The "observable quantities" in the theory of elementary particles. I.)], following Wheeler's 1937 introduction of the S-matrix. [Wheeler, J.A. (1937). *On the Mathematical Description of Light Nuclei by the Method of Resonating Group Structure.*]

Heisenberg's paper attempted to extract from the conceptual structure of the quantum theory of wave fields those terms that were unlikely to be affected by the future change and which would therefore also form a part of the future theory. It avoided the notion of space and time by replacing it with abstract mathematical properties of the S-matrix. The S-matrix related the infinite past to the infinite future in one step, without being decomposable into intermediate steps corresponding to time-slices.

In 1943 Sin-Itiro Tomonaga published a paper in Japanese, of which a translation was published in August 1946, [*On a Relativistically Invariant Formulation of the Quantum Theory of Wave Fields.*] that drew heavily on Dirac (1932) (*Relativistic Quantum Mechanics.*) and Dirac (1933) (*The Lagrangian in Quantum Mechanics.*). It noted that *existing quantum field theory was not relativistic*. Commutation relations referred to points in space at different times. The Schrodinger equation for the vector representing the state of the system was a function of time. Time variable played a different role than space variables. The *probability amplitude* was not *relativistically invariant* in the space-time world. Tomonaga followed Dirac (1932) in *generalizing the notion of probability amplitude as far as was required by the theory of special relativity*. He substituted a four-dimensional form of the commutation relations, then generalized the Schrodinger equation following the Dirac (1933) *many-time* formalism. Tomonaga then introduced his *super many-time theory* in which $[\{H_{12}(P) + h/i\ \partial/\partial C_P\}\ \Psi[C] = 0$ at point P on surface C with infinitely many time variables which represented the local time for each position in the space. This resulted in the *relativistic interaction representation*, in terms of a three-dimensional manifold (space-like "surface") in the four-dimensional space-time world. It was not necessary to also assume time-like surfaces for the variable surface as was required by Dirac. The previous formalism was built up in way too analogous to ordinary *non-relativistic* mechanics. The theory was divided into one section giving the kinematical relations between various quantities at the same instant of time and another section determining the causal relations between quantities at different instants of time; the *commutation relations* belonging to the first section and the *Schrodinger equation* to the second. This way of separating the theory into two sections was very *unrelativistic*, in which the "same instant of time" played a distinct role. The *new formalism consisted of one section giving the laws of behavior of the fields when they were left alone and the other giving the laws determining the deviation from this behavior due to interactions.* This could be carried out *relativistically*. Although the theory was brought into a more satisfactory form no new contents were added. The *divergence difficulties were inherited*, and the fundamental equations admitted only catastrophic solutions *due to non-vanishing zero-point amplitudes of the fields which inhered in the operator $H_{12}(P)$. A more profound modification of theory was required in order to remove this fundamental difficulty.*

In April 1945, Dirac published a paper [*On the Analogy Between Classical and Quantum Mechanics.*] which noted that mathematical methods available for working with *non-commuting quantities* were much weaker than those available for *commuting quantities* owing to the fact that the only functions of *non-commuting variables* that one had been able to define are those expressible algebraically. He showed how this difficulty could be avoided in the case when the *non-commuting quantities* were *observables*, for which it was

31

possible to set up a theory of functions of them (a generalization of the concept known as *well-ordered functions*) of almost the same degree of generality as the usual functions of *commuting variables*. He showed how this theory could be used to make a closer analogy between classical and quantum mechanics, enabling the discussion of trajectories for the motion of a particle in quantum mechanics. A method was given for defining *general functions of non-commuting observables* in quantum mechanics, and developed to provide the formal *probability* for *non-commuting observables* to have numerical values (in general a complex number). This method also enabled the analogy between classical and quantum *contact transformations* to be set up on a more general basis.

The spectrum of hydrogen has a fine structure of the energy levels which according to the *Dirac wave equation* for an electron moving in a Coulomb field was due to the combined effects of *relativistic* variation of mass with velocity and spin-orbit coupling. According to this theory the $2^2S_{1/2}$ state should exactly coincide in energy with the $2^2P_{1/2}$ state which is the lower of the two P states. Previous attempts at measurement had alternated between finding confirmation and discrepancies of as much as eight percent.

In August 1947, Willis Lamb and Robert Retherford published a paper [*Fine Structure of the Hydrogen Atom by a Microwave Method.*] containing the results of their work at the Columbia Radiation Laboratory at Columbia University using a microwave method depending on a novel property of the $2^2S_{1/2}$ level. This showed that, contrary to the *Dirac wave equation*, the $2^2S_{1/2}$ state was higher than the $2^2P_{1/2}$ by about 1000 Mc/sec.

The Lamb and Retherford results showed that the fine structure of the second quantum state of hydrogen did not agree with the *Dirac wave equation*. Schwinger, Weisskopf, and Oppenheimer suggested that this might be due to a *shift of energy levels by interaction of the electron with the radiation field*. This shift came out as infinite in all existing theories and had therefore always been ignored.

Hans Bethe published a response 15 days later [Bethe, H.A. (August, 1947). *The Electromagnetic Shift of Energy Levels.*] which noted that it was possible to identify that the most strongly (linearly) divergent term in the level shift was due to an *electromagnetic mass effect* which must exist for a bound as well as for a free electron, and was therefore already included in the *observed mass* of the electron so should be subtracted. He assumed a *relativistic cut-off* in quantum energies (frequencies) of included atomic states. Then calculation of Lamb shift for hydrogen atom using *non-relativistic* ordinary radiation theory gave a shift of the levels due to *radiation interaction* in close agreement with the observed value, and removed the discrepancy with the Dirac theory. He did not carry out *relativistic* calculations.

In December 1948, Takao Tati and Tomonaga published a paper based on their November 1947 lecture [*A Self-Consistent Subtraction Method in the Quantum Field Theory, I.*] which noted that the procedure previously used by Bloch and Nordsieck [Bloch, F. & Nordsieck, A. (1937). *Note on the Radiation Field of the Electron.*] and then by Pauli and Fierz [Pauli,

W. & Fierz, M. (1938). *Zur Theorie der Emission langwelliger Lichtquanten.* (On the theory of the emission of long-wave light quanta.)] in the treatment of problems with the self-field of an electron first separated the radiation field into a part bound to the electron and a part of unbound photons by means of a canonical transformation, then obtained a term in the Hamiltonian that could be interpreted as the *interaction energy* of electron with the radiation field bound to it. This term though infinite gave rise to a modification of mass of electron so that it could be amalgamated into the mass term in the Hamiltonian for the free electron. They dropped off this term to obtain the *observed mass* by reinterpreting the electron mass as already including it, and obtained the same result as Bethe for level-shift of bound electron in his *non-relativistic* treatment.

Tati and Tomonaga considered it to be desirable to obtain a *relativistic* generalization of this treatment of the self-field of an electron using a canonical transformation. This paper addressed *the field reaction problem when there was no external field to the e^2 approximation.* It did not address the infinity related to *vacuum polarization* effect that occurs in the *relativistic* treatment. This was simply omitted *as it could not be amalgamated into the equation for free radiation.* They started from the *relativistic* formalism of quantum field theory proposed by Tomanaga [August 1946. *On a Relativistically Invariant Formulation of the Quantum Theory of Wave Fields.*] in which the Schrodinger equation with the Hamiltonian for the interaction density between radiation and electron fields was $\{H(P) - i\ \partial/\partial C_P\}\ \Psi[C] = 0$, where $H(P)$ was the *interaction energy density* between radiation and electron fields at the world point P, $\Psi[C]$ was the generalized Schrodinger functional which was a functional of the space-like variable surface C in the four-dimensional world, and $\partial/\partial C_P$ was its partial functional differentiation at the point P, the point P being considered as lying on C. They decomposed the fields into parts oscillating with positive and negative frequencies corresponding to the electron and the positron, and transformed the Schrodinger functional, retaining terms up to the order e^2. They then expanded the integrand into a Fourier integral and defined a 4-vector to obtain the Hamiltonian for the transformed equation. This enabled use of a first order calculation, and allowed processes through intermediate states such as emission and reabsorption of virtual particles (self-energy of an electron) to be treated as direct processes. They then calculated the commutator and rearranged the factors in each term into the correct order. From this they identified (1) terms to represent *interactions between electrons* caused by exchange of virtual photons including ordinary Mø̸ller interaction between two electrons, (2) terms connected with the *self-energy* of an electron, (3) terms responsible for the *scattering of a photon* by a free electron, terms for *creation of a pair* by two photons, (4) variables describing the radiation field only representing modification of the radiation field in vacuo due to the *polarization of the vacuum* that cause an infinite energy level shift of the vacuum itself and an infinite *self-energy* of a photon, and (5) a *mass-modifying term* with a logarithmic diverging quantity representing the electromagnetic mass of an electron. Finally, they assumed that the *mass-modifying term* was already included in the free field equation so no level-shift was caused by the interaction between the electron and radiation.

Their subtraction method was *"self-consistent"* in this sense. It gave a *relativistic* generalization of the transformation that separated the *radiation field* into a field of "unbound" photons and a field of photons bound to the electron, similar to the Hartree method of the *self-consistent* field in which *interaction* between electrons were considered as a perturbation but some part of its effect was already included.

In September, 1948, Ziro Koba and Tomonaga published a paper [*On Radiation Reactions in Collision Processes. I: Application of the "Self-Consistent" Subtraction Method to the Elastic Scattering of an Electron.*], of which a preliminary version was first published in December 1947, which applied Tati and Tomonaga's "self-consistent" subtraction method to the *elastic scattering of an electron by a fixed electrostatic potential*. The formal infinity associated with *mass-modifying term* was attributed to a defect of current theory so *the empirical value was substituted for the theoretical value* (assuming it was already included in the free field equation). This maintained the total Hamilton function describing the interaction of the electron and electromagnetic fields unaltered. It required the *interaction-energy* part of the Hamiltonian to undergo a corresponding change by inclusion of a "counter-self-energy" term ("mass-type" correction to the scattering cross-section), which resulted in a finite value for the *self-energy* of electron in the e^2 approximation. When they applied the new formalism to the *elastic scattering of electron*, the effective cross-section for scattering by a fixed potential in zeroth approximation had a value in good agreement with experiment, but as soon as the reaction of electromagnetic field with the electron was taken into account the correction became infinite. The first infinite term in the modified Hamiltonian was eliminated using the *subtraction hypothesis of positron theory*. The second difficulty disappeared by applying the *self-consistent" subtraction method* using a modified Hamiltonian. The first term can also be eliminated if one interprets the sum of the external potential and its infinite correction due the vacuum polarization effect as the physically observable potential and *substitutes the empirical value* for it, while the interaction part of the Hamiltonian is supplemented by an additional "counter-vacuum polarization" term. *This method by no means gave the real solution of the fundamental difficulty of quantum electrodynamics* but revealed the nature of various diverging terms and reduced them to two quantities - the *self-energy* and the *vacuum polarization*. In this way *it became possible in an unambiguous and consistent manner to treat the field reaction problem without touching the fundamental difficulty by employing the finite empirical values instead of the infinite "theoretical" values for these two quantities*.

All divergences were eliminated by counter-terms in a modified form of the Hamilton function, but this was a simple and particularly favorable case. It was necessary to examine whether this method was still effective in fundamental processes between elementary particles.

In February 1948, Julian Schwinger published a paper [*On quantum-electrodynamics and the magnetic moment of the electron.*] which noted that *electrodynamics unquestionably required revision at ultra-relativistic energies*. It was desirable to isolate those aspects of the current theory that essentially involve high energies and are subject to modification by a more satisfactory theory. He claimed that this goal had been achieved by transforming the Hamiltonian of current *hole theory* to exhibit explicitly the logarithmically divergent *self-energy* of a free electron which arises from the virtual emission and absorption of light quanta. The electromagnetic *self-energy* of a free electron could be ascribed to an *electromagnetic mass* which must be added to the mechanical mass of the electron. The new Hamiltonian involved the *experimental electron mass* rather than the *unobservable mechanical mass*; the electron then interacted with the radiation field only in the presence of an external field such that only an accelerated electron could emit or absorb a light quantum. The *interaction energy* of an electron with an external field was now subject to a *finite* radiative correction. *Polarization of the vacuum* still produced a logarithmically divergent term proportional to the *interaction energy* of an electron in an external field, which was equivalent to altering the value of the *electron charge* by a constant factor with only the final value being identified with the experimental charge. This resulted in the interaction between matter and radiation producing a *renormalization* of the electron charge and mass, with all divergences contained in the *renormalization* factors. The radiative correction for the energy of an electron in an external magnetic field corresponded to an *additional magnetic moment associated with electron spin* of magnitude $\delta\mu/\mu = (\frac{1}{2}\pi)e^2/hc = 0.001162$. The experimental measurements on the hyperfine splitting of the ground states of atomic hydrogen and deuterium were larger than expected from directly measured nuclear moments. It was found that the additional *electron spin magnetic moment* accounted for measured hydrogen and deuterium *hyperfine structures* to be $\delta\mu/\mu = 0.00126$ and $\delta\mu/\mu = 0.00131$ respectively, with these discrepancies accounted for by additional spin magnetic moment to the electron of $\delta\mu/\mu = 0.0018 \pm 0.00003$. *The values yielded by this relativistic calculation of the Lamb shift differed only slightly from those conjectured by Bethe on the basis of non-relativistic calculation and were in good accord with experiment.*

In December, 1948, Koba and Gyo Takeda published a paper, received May, 1948, [*Radiation Reaction in Collision Process, II: Radiative Corrections for Compton Scattering.*] which applied this method to calculate e^2-corrections to *Klein-Nishina formula for Compton scattering* using the *perturbation method*. It noted two types of diverging terms, one related to *polarization of vacuum* and other to *self-energy* of electron, and showed that it is sufficient to add two new terms to Hamiltonian function to cancel out these infinite terms. The *self-energy* logarithmic divergence was eliminated by introducing a new term into interaction Hamiltonian derived in a plausible way as a "counter-term" which compensates the change in the free field Hamiltonian and conserves the total Hamiltonian unaltered. One of the *vacuum polarization* terms could be treated in similar way, while other term could not be foisted into the theory without radically changing the Maxwell equation for the free electromagnetic field. This was *far from the final settlement of the fundamental problem in theory of elementary particles.*

In June 1948, Koba and Takeda published a letter [*Radiative Corrections in e² for an Arbitrary Process Involving Electrons. Positrons, and Light Quanta.*] which investigated further the *radiation correction in collision processes between electrons, positrons and photons* in a general manner and confirmed that a modified Hamiltonian was sufficient to eliminate the divergence difficulty *in the case of the first radiative correction.* They *introduced the "transition diagram method"* [similar to the Stueckelberg diagrams [Stueckelberg, E.C.G. (1941). *Remarque apropos de la creation de paires de particules en theorie de relativite.*] later introduced by Feynman at Ponoco in spring 1948 and published in September 1949 [*The Theory of Positrons.*], and subsequently renamed *Feynman diagrams*] *to analyze the complicated connection between initial and final states through a number of intermediate ones. Electrons* and *positrons* were expressed as *world lines* in *momentum space*, while *emission* and *absorption* of *photons* were described as *"leaps"* of these *world lines.* The diagram acquired two additional *leaps* (the *emission* and *reabsorption* of a *virtual photon*) when *radiative corrections* were taken into account. There were two distinct ways of attaching them, existing *world lines* gained new *leaps* or additional closed *world line* were introduced and coupled with existing ones through *virtual photon.* The first case implied *"mass-type"* or *"self-energy-type"* divergences when the *emission* and *reabsorption* of the *virtual photon* take place in succession, which can be eliminated by a *counter-self-energy term.* The second case represented *vacuum polarization* corrections, which can be subdivided according to the number of *leaps* in additional closed *world lines*, which afford quadratic and logarithmic divergences that are cancelled by the "*counter-terms*" and found not to contribute to the divergence.

In April 1948 Richard Feynman published a paper [Feynman, R.P., (1948). *Space-Time Approach to Non-Relativistic Quantum Mechanics.*] which described a third formulation of *non-relativistic* quantum theory in addition to the *differential equation of Schroedinger* and the *matrix algebra of Heisenberg.* This was the *path integral formulation*, utilizing the *action* principle as suggested in Dirac (1933) [*The Lagrangian in Quantum Mechanics.*] and Dirac (1945) [*On the Analogy Between Classical and Quantum Mechanics.*]. In quantum mechanics the probability of an event which can happen in several different ways is the absolute square of a sum of the complex contributions, one from each alternative way, $\varphi_{ac} = \sum_b \varphi_{ab}\varphi_{bc}$ where φ_{ab}, φ_{bc}, φ_{ac} are complex numbers such that $P_{ab} = |\varphi_{ab}|^2$, $P_{bc} = |\varphi_{bc}|^2$, and $P^q_{ac} = |\varphi_{ac}|^2$, where P_{ab} is the probability that if measurement A gave the result a, then measurement B will give the result b, and P^q_{ac} is the quantum mechanical probability that a measurement of C results in c when it follows a measurement of A giving a. The probability that a particle will be found to have a path lying somewhere within a region of space time is the absolute square of a sum of contributions, one from each path in the region. The contribution from a single path was postulated to be an exponential whose (imaginary) phase is the classical *action* for the path in question where *action* refers to time integral of Lagrangian along a path. This was restricted to a finite time interval. *The probability amplitude for a space-time path was associated with the entire motion of the particle as a function of time rather than with the position of the particle at a particular time*. The total contribution from all paths reaching x, t from the past was the wave function $\psi(x, t)$. This was shown to satisfy Schroedinger's equation. Postulates were established that described *non-relativistic quantum mechanics neglecting spin*, which were mathematically equivalent to Heisenberg and Schroedinger formulations. There were no fundamentally new results, and this formulation *suffered serious drawbacks*. It required an unnatural and cumbersome division of the time interval, and was not formulated so that it was physically obvious that it was invariant under unitary transformations, though improvements could be made through the use of the notation and concepts of mathematics of functionals.

Quantum electrodynamics was built from a classical counterpart that already contained many difficulties which remained upon quantization. It had been hoped that if a classical electrodynamics could be devised which did not contain the difficulty of *infinite self-energy*, and this theory could be quantized, then the problem of a self-consistent quantum electrodynamics would be solved. Previous attempts to address the problem of *infinite self-energy* that resulted from assuming point electron in *relativistic* theory were met with considerable difficulties when attempts were made to quantize them.

In October 1948, Feynman published a paper [*A Relativistic Cut-Off for Classical Electrodynamics.*] which described *a consistent classical theory which he believed could be quantized*. The *potential* at a point in space at a given time depended on the charge at a distance r from the point at a time previous by t = r (taking the speed of light as unity). *Relativistically*, interaction occurred between events whose four-dimensional interval, s, defined by $s^2 = t^2 - r^2$, vanished. It was formulated in terms of *action at a distance*. This resulted in an infinite *action* of a *point electron* on itself. This theory was essentially that of Friedrich Bopp (1942). It modified this idea by assuming that substantial interaction exists as long as the interval s was time-like and less than some small length, *a*, of order of the electron radius. This reduced the infinite *self-energy* to a finite value for accelerations which were not extreme, in which the *action* of an electron on itself appeared as *electromagnetic mass*. It satisfied Maxwell's equations; not the usual *retarded* solution for which there was no *self-force* but the half the *retarded* plus half the *advanced* solution in Wheeler and Feynman (1945) [*Interaction with the Absorber as the Mechanism of Radiation.*]. The effect of the modification was to change slightly the field of one particle on another when they are very close, and to add a *self-force*. Feynman concluded that there was *little reason to believe that the ideas used here to solve the divergences of classical electrodynamics would prove fruitful for quantum electrodynamics*.

In November 1948, Feynman published a paper [*Relativistic Cut-Off for Quantum Electrodynamics.*] which described a model based on the quantization of the classical theory for which all quantities automatically come out finite described in his previous paper. This contained an arbitrary function on which numerical results depend. The only term that depended significantly (logarithmically) on the cut-off frequency was the *self-energy* which could be used to renormalize the electron mass. The remaining terms were nearly independent of the function. This model applied only to results for processes in which virtual quanta were emitted and absorbed. *Terms representing processes involving a pair production followed by annihilation of the same pair were infinite and not made convergent by this scheme.* Problems of *permanent emission* and the position of *positron theory* still need to be addressed. *This paper may be looked upon as presenting an arbitrary rule to cut off at high frequencies in a relativistically invariant manner the otherwise divergent integrals appearing in quantum field theories.* It produced finite invariant *self-energy* for a free electron, but the problem of *polarization of the vacuum* was not solved. An alternative cut-off procedure which eliminated high frequency intermediate states offered to solve the *vacuum polarization* problems as well.

Lack of convergence in current formulations of *quantum electrodynamics* indicated that a revision of electrodynamic concepts at *ultra-relativistic* energies was necessary. Elementary phenomenon in which divergences occurred as a result of virtual transitions involving particles with unlimited energy were *polarization of the vacuum* and *self-energy of the electron* which expressed *the interaction of the electromagnetic and matter fields with their own vacuum fluctuations.* This altered the constants characterizing the properties of the individual fields and their mutual coupling by infinite factors. The question was whether all divergencies could be isolated in such unobservable *renormalization* factors.

In November 1948, Schwinger published a paper [*Quantum Electrodynamics. I. A Covariant Formulation.*], which was occupied with *the formulation of a completely covariant electrodynamics*. He asserted that manifest covariance with respect to Lorentz and gauge transformations was essential in a divergent theory. Customary *canonical commutation relations* failed to exhibit the desired covariance since they referred to field variables at equal times and different points of space. They could be put in a covariant form by replacing the four-dimensional surface t = const. by a space-like surface. This offered the advantage over the Schrodinger representation, in which all operators refer to the same time, by providing a distinct separation between *kinematical* and *dynamical* aspects. A formulation that retained the evident covariance of the Heisenberg representation but offered something akin to the Schrodinger representation could be based on the distinction between the properties of *non-interacting fields*, and the effects of *coupling between fields*. In the second section, he constructed a *canonical transformation* that changed the *field equations* in the *Heisenberg representation* into those of *non-interacting fields*. He added *supplementary condition* restricting the admissible states of the system and the *commutation relations* to the *equations of motion* to obtain a description of the coupling between fields in terms of a varying *state vector*. Then it was a simple matter to evaluate commutators of *field* quantities at arbitrary *space-time* points. He thus obtained an obviously covariant and practical form of quantum electrodynamics expressed in a mixed Heisenberg-Schrodinger representation, called the *interaction representation*. The third section discussed the *covariant* elimination of the longitudinal field in which the customary distinction between longitudinal and transverse fields was replaced by a suitable *covariant* definition. The fourth section described collision processes in terms of an invariant *collision operator*, which was the unitary operator that determined the over-all change in state of a system as the result of interaction. He noted that a *second paper* would treat the problems of the electron and photon *self-energy* together with the *polarization of the vacuum*, and a *third paper* was concerned with the determination of the *radiative corrections* to the properties of an electron and the comparison with experiment. This was not addressed in that paper; it stated that "*radiative corrections to energy levels* will be treated in the next paper of the series" but *this did not appear nor are there any references to it.*

The recent and independent formulations of quantum electrodynamics by Tomonaga, Schwinger, and Feynman have made two notable advances, the foundations and applications of the theory have been simplified by being presented in *a completely relativistic way* and the *divergence difficulties have been at least partially overcome*. The advantages of the Feynman formulation were simplicity and ease of application while those of Tomonaga-Schwinger were generality and theoretical completeness,

In February 1949, Freeman Dyson, at age 24, published a paper [Dyson, F.J. (1949). *The Radiation Theories of Tomonaga, Schwinger, and Feynman.*] which presented a unified development of quantum electrodynamics embodying the main features of the Tomonaga-Schwinger and the Feynman radiation theories. It aimed to show how the Schwinger theory could be applied to specific problems in such a way as to incorporate the ideas of Feynman. The emphasis was on the application of the theory. The main results were general formulas from which radiative reactions on the motions of electrons could be calculated. It divided *energy-density* into two parts, the *energy of interaction* of two fields with each other and the energy produced by external forces. *Interaction energy* alone was treated as a perturbation. Important results of the paper were the *equation of motion* ihc[$\partial\Omega/\partial\sigma(x_0)$] = {S($\sigma$)}$^{-1}$H^e(x$_0$)S($\sigma$)$\Omega$ for the *state vector* $\Omega(\sigma)$, and the interpretation of the *state vector* Ω. It simplified the Schwinger theory for using it for calculations, and demonstrated the equivalence of the theories within their common domain of applicability. In the *Schwinger theory* the aim was to calculate the matrix elements of the "*effective external potential energy*" between *states* specified by their *state vectors*. In the *Feynman theory* the basic principle was to "preserve symmetry between past and future" so the matrix elements of the operator were evaluated in a "*mixed representation*" in which the matrix elements were calculated between an *initial state* specified by its *state vector* and a *final state* specified by its *state vector*. A graph corresponding to a particular matrix element was used not merely as an aid to calculation but as a picture of the physical process which gave rise to that matrix element. Dyson derived fundamental formulas for the operator in the *equation of motion* for the *state vector* $\Omega(\sigma)$ which represented the interaction of a physical particle with an external field for both the Schwinger and the Feynman theories, and a set of rules by which matrix element of Feynman operator might be written down in a form suitable for numerical evaluation. He showed the equivalence of two theories, and developed graphical representations of the matrix elements. He noted that *the theory as a whole could not be put into a finally satisfactory form so long as divergencies occurred in it however skillfully these divergencies were circumvented*. The present treatment should be regarded as justified by its success in applications rather than by its theoretical derivation. The *paper suffered from a series of significant errors*.

In March and June 1949, Hiroshi Fukuda, Yoneji Miyamoto, and Tomonaga published a sequel [*A Self-Consistent Subtraction Method in the Quantum Field Theory. II-1 and II-2.*] to Tati and Tomonaga (1948) [*A Self-Consistent Subtraction Method in the Quantum Field Theory, I.*], which addressed the case when an *external field is present*. The unbound fields no longer propagate freely and transitions took place. The electron wave was able to change its state of propagation not only elastically by the external field but also by emitting or absorbing unbound photons. Thus, *an interaction appeared between an electron and radiation* which were free from interaction with each other in the absence of the external field. This interaction caused a radiation reaction upon the electron so that the motion of the electron would be modified. *In this paper they gave an example of how one could calculate this reaction and obtained a finite result as the consequence of the subtraction procedure*. They were able to obtain a *finite radiative level-shift* of a bound electron in the external field and a *finite e^2-correction* to the *scattering cross section*, and estimated the Lamb shift for hydrogen atom as 10^{76} Mcycles by adding additional terms to Bethe's value. They thus successfully obtained finite answers for these field reaction problems of the magnitude agreeing with experimental results. But they also noted "*we must nevertheless confess that the calculation carried out in this paper is still unsatisfactory because we had to make a non-relativistic treatment in the evaluation of the effective energies … Our work is therefore only of a provisionary character. … It is still problematic whether this procedure corresponds to the correct prescription*". The new formalism of quantum field theory was more satisfactory from the stand-point of *relativistic invariance* as it was formulated in terms of *invariant space-time* concepts, but it employed *canonical formalism* in which the time variable was unnecessarily distinguished from the space variables.

In September 1949, Suteo Kanesawa and Koba published a paper [Kanesawa, S. & Koba, Z. (1949). *A Remark on Relativistically Invariant Formulation of the Quantum Field Theory.*] which showed that *the same formalism could be reached without referring to the canonical formulation*. In the current theory the *interaction Hamilton density* coincided with the *interaction Lagrange density*, except the sign, if the surface-dependent part was neglected. This suggested the idea of connecting the *generalized interaction Lagrange density* directly to generalized Schrodinger equation. The guiding principle was the integrability of the Tomonaga-Schwinger equation. This paper showed how to apply the new method to the general case when a canonical description was impossible. *Problematic aspects of theory such as universal length or ultraviolet divergencies were not addressed.* The aim of the paper lay in demonstrating the possibility of including more general kinds of interactions in field theory other than those which allowed a canonical description.

The initial calculation of the correct Lamb shift by Bethe (1947) [*The Electromagnetic Shift of Energy Levels.*] was *non-relativistic*. Various *relativistic* calculations followed the next year. A comedy of errors ensued: *both Feynman and Schwinger made an incorrect patch between hard and soft photon processes, and so obtained identical, but incorrect, predictions for the Lamb shift,* and the weight of their reputations delayed the publication of the correct, if pedestrian, calculations by Kroll & Lamb and French & Weisskopf until February 1949.

In February 1949, Norman Kroll and Willis Lamb published a paper [Kroll, N.M. & Lamb, Jr., W.E. (1949). *On the self-energy of a bound electron.*] which contained their calculation of the electromagnetic shift of energy levels of a bound electron based on the usual formulation of *relativistic* quantum electrodynamics and *positron theory*. This gave 10^{52} megacycles per second for the $^{2}2S_{1/2} - ^{2}2P_{1/2}$ shift in hydrogen in close agreement with the *non-relativistic* calculation of 10^{40} megacycles per second by Bethe.

In April 1949, James Bruce French and Victor Weisskopf published a paper [*The Electromagnetic Shift of Energy Levels.*] which contained their *relativistic* calculation of Lamb shift using the conventional form of perturbation theory and removing the infinite *self-energy* of the electron by subtracting a "mass operator" from the Hamiltonian. This gave 10^{51} megacycles per second for $2s_{1/2} - 2p_{1/2}$ separation in hydrogen, compared with the *non-relativistic* calculation of 10^{40} megacycles per second by Bethe, and the *relativistic* calculation of 10^{52} megacycles per second by Kroll & Lamb. The results of the different calculations suggested that they were not dependent on whether they are *relativistic* or *non-relativistic*.

In February 1949, Schwinger published a paper [*Quantum Electrodynamics. II. Vacuum Polarization and Self-Energy.*] which applied the *interaction representation* to the *polarization of the vacuum* and the *self-energies of the electron and photon*. In the *first section* the vacuum of the non-interacting *electromagnetic* and *matter* fields was *covariantly* defined as the *state* for which the eigenvalue of an arbitrary time-like component of the *energy-momentum four-vector* was an absolute minimum. The covariant decomposition of field operators into positive and negative frequency components was introduced to characterize the *vacuum state vector*. He showed that the *state vector for the electromagnetic vacuum* was annihilated by the positive frequency part of transverse four-vector potential and *state vector for matter vacuum* was annihilated by positive frequency part of Dirac spinor and its charge conjugate. These properties of *vacuum state vector* were employed in the calculation of the *vacuum expectation* values of quadratic field quantities, specifically the *energy-momentum tensors* of the independent *electromagnetic* and *matter fields* and the *current four-vector*. It was inferred that the *electromagnetic energy-momentum tensor* and *current vector* must vanish in the vacuum, while the *matter field energy-momentum tensor* vanished in the vacuum only by the addition of a suitable multiple of the unit tensor. The *second section* treated the *induction of a current in the vacuum by an external electromagnetic field*. It was supposed that the *external electromagnetic field* did not produce actual electron-positron pairs, and considered only the phenomenon of *virtual pair creation*. This restriction was introduced by requiring that the establishment and subsequent removal of the external field produced no net change in state for the *matter field*. *He demonstrated that the induced current at a given space-time point involved the external current in the vicinity of that point and not the electromagnetic potentials*. This *gauge invariant* result showed that a light wave propagating at remote distances from its source induced no current in the vacuum and was therefore undisturbed in its passage through space. This indicated an *absence of a light quantum self-energy effect*. The *current* induced at a point consisted of two parts, a logarithmically divergent multiple of the *external current* at that point, *which produced an unobservable renormalization of charge*, and a more involved finite contribution, which is the physically significant *induced current*. *The third section considered the modification of matter field properties arising from the interaction with the vacuum fluctuations of the electromagnetic field*. This analysis was carried out with two alternative formulations, one employing the complete *electromagnetic potential* together with a *supplementary condition*, the other using the *transverse potential* with the variables of the *supplementary condition* eliminated. It was noted that no real processes were produced by the first order coupling between the fields. Alternative *equations of motion* for the *state vector* were constructed from which the first order interaction term had been eliminated and replaced by the *second order coupling* which it generated. The latter included the *self-action* of individual particles and light quanta, the *interaction* of different particles, and a *coupling* between particles and light quanta which produced such effects as Compton scattering and two quantum pair annihilation. It was concluded from a comparison of the alternative procedures that, *for the treatment of virtual light quantum processes, the separate consideration of longitudinal and transverse fields was an inadvisable complication*. The light quantum the *self-energy* term was shown to vanish, while that for a particle had the form for a change in *proper mass* but was logarithmically divergent in agreement with previous calculations. The identification of *self-energy* effect with a change in *proper mass* was confirmed by removing this term from the state vector *equation of motion*, which altered the *matter field equations of motion* in the expected manner. It was verified that *the energy and momentum modifications produced by self-interaction effects were entirely accounted for by the addition of the electromagnetic proper mass to the mechanical proper mass—an unobservable mass renormalization*. An appendix was devoted to the construction of several invariant functions associated with the *electromagnetic* and *matter* fields.

In June 1949, Dyson published a paper [*The S Matrix in Quantum Electrodynamics.*] in which the covariant *quantum electrodynamics* of Tomonaga, Schwinger and Feynman was used as basis for a general treatment of *scattering problems* involving electrons, positrons, and photons. It addressed the relation between Schwinger and Feynman theories when the restriction to one-electron problems was removed. In these more general circumstances, the two theories appeared as complementary rather than identical. *The Feynman method was essentially a set of rules for the calculation of the elements of the Heisenberg S matrix* corresponding to any physical process, and could be applied directly to all kinds of scattering problems. *The Schwinger method evaluated radiative corrections by exhibiting them as extra terms appearing in the Schrodinger equation.* The paper showed the practical usefulness of the S matrix as a connecting link between the Feynman technique of calculation and Hamiltonian formulation of quantum electrodynamics. The *Feynman radiation theory* provided a set of rules for the calculation of matrix elements between *states* composed of any number of ingoing and outgoing free particles. It was thus an S matrix theory. The paper showed that *scattering processes*, including the creation and annihilation of particles, were completely described by Heisenberg's S matrix. The elements of this matrix were calculated by a consistent use of perturbation theory to any desired order. Detailed rules were given for carrying out such calculations, and divergences arising from higher order radiative corrections were removed from the S matrix by consistent use of *mass and charge renormalization*. The operators so calculated were divergence-free, *the divergent parts at every stage of the calculation being explicitly dropped after being separated from the finite parts.* This involved *extensive manipulations of infinite quantities* and had to be justified *a posteriori* by the fact that they ultimately lead to a clear separation of finite from infinite expressions. Dyson claimed that such an *a posteriori* justification of dubious manipulations was an inevitable feature of any theory which aims to extract meaningful results from not completely consistent premises. The perturbation theory of this paper was *applicable only to a restricted class of problems* when not only the radiation interaction but also the external potential was small enough to be treated as a perturbation. *It did not give a satisfactory approximation either in problems involving bound states or in scattering problems at low energies.* In other situations, the Schwinger theory would have to be used in its original form. The *problems of extending this treatment to include bound-state phenomena and of proving convergence of the theory as the order of perturbation itself tends to infinity was not addressed.* This analysis suggested that *the divergences of electrodynamics were directly attributable to the fact that the Hamiltonian formalism was based upon an idealized conception of measurability.* Now it was no longer a compelling necessity for a future theory to abandon some essential features of the present electrodynamics. The present electrodynamics was certainly incomplete, but was no longer certainly incorrect.

The paper published by Richard Eden in September 1949 [*Heisenberg's S Matrix for a System of Many Particles.*] was included as a vignette. Dr. Eden provided lectures in Nuclear Physics and Quantum Theory to the author in 1963 as part of his M.A. Cantab. degree. Eden's thesis advisors were Dirac and Heisenberg. The author attended a joint lecture by Heisenberg and Dirac on May 23, 1963, (both speaking whilst writing and erasing on dueling pairs of blackboards at the same time) at the old Cavendish Laboratory.

In September 1949, Feynman published a paper [Feynman, R.P. (1949). *The Theory of Positrons.*] which was the first of a set of papers dealing with the solution of problems in *quantum electrodynamics* in which the main principle was to deal directly with *solutions* of the Hamiltonian differential equations rather than with equations themselves. This paper *analyzed the motion of electrons and positrons in given external potentials, neglecting interaction*, by dealing directly with the *solutions* of the Hamiltonian time differential equations rather than with the equations. A second paper below considers interactions (*quantum electrodynamics*). The problem of charges in a fixed potential was usually treated by method of second quantization of the electron field using the *theory of holes*. Here Dirac's "hole theory" was replaced by reinterpreting the *solutions* of the Dirac equation. The results were simplified by *following the charge rather than the particles* because *the number of particles is not conserved whereas charge is conserved*. In the approximation of classical *relativistic* theory, the creation of an electron pair (electron and positron*)* was represented by the start of two world lines from the point of creation. *Following the charge rather than the particles corresponds to considering the continuous world line as a whole rather than breaking it up into its pieces.* This over-all *space-time* point of view led to considerable simplification in many problems. Quantum mechanically the direction of the world lines was replaced by the direction of propagation of waves. This was quite different from Hamiltonian method which considered the future as developing continuously from the past. In a scattering problem this over-all view of the complete scattering process was similar to the *S-matrix view-point of Heisenberg*. The temporal order of events during scattering which was analyzed in such detail by the Hamiltonian differential equation was irrelevant. This development stemmed from the idea that in *non-relativistic* quantum mechanics the *amplitude* for a given process could be considered as the sum of the *amplitude* for each *space-time* path available. In the *relativistic* case the restriction that the paths must proceed always in one direction in time was removed. The results were more easily understood from the more familiar physical viewpoint of *scattered waves* used in this paper. After the equations were worked out physically the proof of the equivalence to the *second quantization theory* was found. The solution in terms of boundary conditions on *wave function* contained all possibilities of pair formation and annihilation together with the ordinary scattering processes. Negative energy states appeared in space-time as waves traveling away from the external potential backwards in time (as suggested in Stueckelberg [Stueckelberg, E.C.G. (1941). *Remarque apropos de la creation de paires de particules en theorie de relativite.*]). Such a wave corresponded to a positron approaching the potential and annihilating the electron. A particle moving forward in time (electron) in a potential might be scattered forward in time (ordinary scattering) or backward (pair annihilation). When moving backward (positron) it might be scattered backward in time (positron scattering) or forward (pair production). The *amplitude* for a transition from an initial to a final state could be analyzed to any order in the *potential* by considering it to undergo a sequence of such scatterings. The *amplitude* for a process involving many such particles was the product of the *transition amplitudes* for each particle. Vacuum problems did not arise for charges which did not interact with one another. The equivalence to the *theory of holes* in second quantization was demonstrated in an appendix.

Electrodynamics was viewed as direct *interaction* at a distance between *charges* rather than the behavior of a *field* (Maxwell's equations). The *field point of view*, which separated the production and absorption of light, was most practical for problems involving *real quanta* while the *interaction view* was best for a discussion of *virtual quanta* when dealing with

44

close collisions of particles or their actions on themselves. The Hamiltonian method was not well adapted to represent direct *action at a distance* between *charges* because action was delayed. This forced the use of the field viewpoint rather than the interaction viewpoint. For *collisions* it was much easier to treat the process as a whole. The effects of longitudinal and transverse waves could be combined.

In a continuation of the previous paper in the same issue [Feynman, R.P. (1949). *Space-Time Approach to Quantum Electrodynamics.*] Feynman applied the same technique to include *interactions* and, in that way, to express in simple terms of the solution of problems in *quantum electrodynamics* rather than the differential equations from which they came. (1) It was shown that *a considerable simplification could be attained by writing down matrix elements for complex processes in electrodynamics*. A physical point of view was available which permitted them to be written down directly. The simplification resulted from the fact that previous methods separated into individual terms processes that were closely related physically. (2) *Electrodynamics was modified by altering the interaction of electrons at short distances*. All matrix elements were now finite with the exception of those relating to *vacuum polarization*. The latter were evaluated in a manner suggested by Pauli and Bethe to give finite results. Phenomena directly observable were insensitive to the details of the modification. Feynman began with the solution in *space* and *time* of the Schrodinger equation for particles interacting instantaneously. The solution was expressed in terms of *matrix elements, derived from the Lagrangian form of quantum mechanics*. It was then modified in accordance with the requirements of the Dirac equation and the phenomenon of pair creation. This was made easier by reinterpreting the *theory of holes*. This was generalized to *delayed interactions* of *relativistic* electrons. For practical calculations expressions were developed in a power series of $e^2/\hbar c$. The *derivation would appear in a separate paper*. By forsaking Hamiltonian method, the wedding of *relativity* and *quantum mechanics* was accomplished most naturally. The *relativistic* invariance became self-evident but the *matrix elements for complex processes* and the *self-energy diverged*. Then it was possible to see how the matrix elements could be written down directly. Feynman then introduced a modification in the interaction between charges at short distances - *he assumed that substantial interaction existed as long as the four-dimensional interval was time-like and less than some small length of order of the electron radius* [Feynman (1948). *A Relativistic Cut-Off for Classical Electrodynamics.*]. *Convergence factors* were then introduced such that the integrals with their convergence factors now converged. The *self-energy* was now convergent, and corresponded to a correction to the electron mass. *All matrix elements were now finite* with exception of those relating to *vacuum polarization*, but *a strict physical basis for rules of convergence was not known*. After *mass and charge renormalization*, the results were equivalent to those of Schwinger in which the terms corresponding to corrections in mass and charge were identified and removed from the expressions for real processes. *Although in the limit the two methods agreed neither method appeared to be completely satisfactory theoretically.* The practical advantage of the new method was that ambiguities could be more easily resolved by direct calculation of otherwise divergent integrals. A complete method was therefore available for calculations of all processes involving electrons and photons. It showed how matrix element for any process could be written down directly in *Feynman diagrams*. This paper included the first published "*Feynman diagram*". *This attempt to find a consistent modification of quantum electrodynamics was incomplete.* It was *not at all clear that the convergence factors did not upset the physical consistency of the theory.*

A covariant form of *quantum electrodynamics* had been developed and applied by Schwinger in the previous articles of this series [Schwinger, J. (November, 1948). *Quantum Electrodynamics. I. A Covariant Formulation*; and (February, 1949). *Quantum Electrodynamics. II. Vacuum Polarization and Self-Energy.*] to two elementary phenomena that are produced by the *vacuum fluctuations of the electromagnetic field*. These applications were the *polarization of the vacuum* expressing the modifications in the properties of an electromagnetic field arising from its interaction with the *matter field* vacuum fluctuations, and the *electromagnetic mass of the electron* embodying the corrections to the mechanical properties of the *matter field* in its single particle aspect. In these problems *the divergences that mar the theory were found to be concealed in unobservable charge and mass renormalization factors.* The previous paper [*Quantum Electrodynamics. II. Vacuum Polarization and Self-Energy.*] was confined to consideration of *vacuum polarization* produced by the field of a prescribed *current* distribution.

In September 1949, Schwinger published a paper [Schwinger, J. (September, 1949). *Quantum Electrodynamics. III. The Electromagnetic Properties of the Electron—Radiative Corrections to Scattering.*] in which he now *considered how the induction of a current in a vacuum by an electron resulted in an alteration in its electromagnetic properties* revealed by scattering in Coulomb field and energy level displacements. *This paper was concerned with the computation of the second-order corrections to the current operator as modified by the coupling with the vacuum electromagnetic field and its application to electron scattering* by a Coulomb field. It applied canonical transformation to *renormalize the electron mass*. A correction to the *current operator* produced by coupling with electromagnetic field was developed in power series, of which the first- and second-order terms were retained. Second-order modifications in the *current operator* were obtained which were of same general nature as the previously treated *vacuum polarization current*, apart from a contribution in the form of a *dipole current*. The latter implied a fractional increase of $\alpha/2\pi$ in the *spin magnetic moment* of electron. The only flaw in the second-order current correction was a *logarithmic divergence attributable to an infra-red catastrophe*. In the presence of an *external field* the first-order *current* correction introduced a compensating divergence. Thus, the second-order corrections to the electromagnetic properties of a particle could not be completely stated without regard for the manner of exhibiting them by an *external field*. Accordingly, in the second section, the interaction of three systems - the *matter field*, the *electromagnetic field*, and a given *current distribution* - was considered. It was shown that this could be described in terms of an *external potential* coupled to the *current* operator as modified by the interaction with the

46

vacuum electromagnetic field. This was applied to *scattering of an electron* by an *external field* which was regarded as a small perturbation. It was convenient to calculate the total rate at which collisions occur and then identify the *cross sections* for individual events. The correction to the *cross section* for radiation-less scattering was determined by the second-order correction to the *current* operator. Scattering that was accompanied by a single quantum emission was a consequence of the first-order *current* correction. The final object of calculation was the *differential cross section* for scattering through a given angle with a prescribed maximum energy loss which was completely free of divergences. An *Appendix* was devoted to an alternative treatment of the *polarization of the vacuum* by an external field. It was noted that *radiative corrections to energy levels would be treated in the next paper of the series [but this did not appear]*.

In two previous papers [Feynman, R.P. (1949). *The Theory of Positrons.* and Feynman, R.P. (1949). *Space-Time Approach to Quantum Electrodynamics.*] Feynman gave rules for the calculation of the matrix element for any process in electrodynamics. No complete proof of the equivalence of these rules to conventional electrodynamics was given, nor was a closed expression given valid to all orders in $e^2/\hbar c$.

In November 1950, Feynman published a paper [Feynman, R.P. (1950). *Mathematical Formulation of the Quantum Theory of Electromagnetic Interaction.*] which addressed these formal omissions, giving the derivations of the formulas of Feynman (1949). [*Space-Time Approach to Quantum Electrodynamics.*] by means of the form of quantum mechanics given in Feynman (1948). [*Space-Time Approach to Non-Relativistic Quantum Mechanics.*]. The derivation of the rules used the Lagrangian form of quantum mechanics, which permitted the motion of any part to be solved first and the results to be used in the solution of the motion of other parts. The electromagnetic field is a simple system. The *interaction of matter (electrons and positrons) and the field were analyzed by first solving for the behavior of the field in terms of the coordinates of the matter.* Integration over field oscillator coordinates eliminated field variables from the *equations of motion* of electrons. He then addressed the behavior of electrons. In this way, all of the rules given in the second paper were derived. This was *restricted to cases in which the particle's motion was non-relativistic* but Feynman claimed that the transition of the final formulas to the relativistic case was direct and the proof could have been kept relativistic throughout. The *generalized formulation was unsatisfactory because for situations of importance it gave divergent results.* Problems of divergences were not discussed. The *theory assumed that substantial interaction existed as long as the interval was time-like and less than some small length of the order of the electron radius.*

In October 1951, Feynman published another paper [Feynman, R.P. (1951). *An Operator Calculus Having Applications in Quantum Electrodynamics.*] which suggested an alteration in the mathematical notation for handling operators. The new notation permitted a considerable increase in the ease of manipulation of complicated expressions involving operators. *No new results were obtained in this way. The mathematics was not completely satisfactory and no attempt had been made to maintain mathematical rigor*. Feynman believed that to put these methods on a rigorous basis might be quite a difficult task, beyond the abilities of the author.

In October 1962, Dirac published a paper [*The Conditions for a Quantum Field Theory to be Relativistic.*] which noted that a quantum field theory in agreement with *special relativity* could be built up from the infinitesimal operators of translation and rotation. These operators were expressible in terms of a momentum density and an energy density. The momentum density was determined by the geometrical properties of the fields concerned. The energy density had to satisfy commutation relations for which certain conditions hold.

On December 11, 1966, Tomonaga, received the 1965 Nobel Prize in Physics, together with Schwinger and Feynman, "for their fundamental work in *quantum electrodynamics*, with deep-ploughing consequences for the physics of elementary particles".

Tomonaga received the prize for his reformulation of the *relativistic* theory for the interaction between charged particles and electromagnetic fields as a consequence of the observation of the Lamb shift in 1947, in which the supposed single energy level within a hydrogen atom was instead proven to be two similar levels. Tomonaga solved this problem in 1948, when he was 42 years old, through a "*renormalization*" and thereby contributed to a new quantum electrodynamics.

Tomonaga's Nobel Prize lecture [Tomonaga, S. (May 6, 1966). *Nobel Lecture. Development of Quantum Electrodynamics - 1966: Personal recollections.*] described the evolution of his work and that of others since 1932 when he started his research career up until 1948.

Feynman was awarded the prize for contributing to creating a new quantum electrodynamics by introducing Feynman diagrams in 1948, when he was 30 years old. These were graphic representations of various interactions between different particles which facilitated the calculation of interaction probabilities.

Feynman's Nobel Prize lecture [Feynman, R. P. (December 11, 1965). *Nobel Lecture. The Development of the Space-Time View of Quantum Electrodynamics.*] described the sequence of ideas which occurred and by which he finally came out the other end with an unsolved problem for which he ultimately received the Nobel prize. He represented conventional electrodynamics with *retarded interaction*, not his half-advanced and half-retarded theory. The *action* expression was not used. The idea that *charges* do not act on themselves was abandoned. The *path-integral formulation* of quantum mechanics was useful for guessing at final expressions and at formulating the general theory of electrodynamics in new ways though was not absolutely necessary. The rest of his work was simply to improve the techniques then available for calculations, *making diagrams to help analyze perturbation theory quicker*. He concluded: "*I don't think we have a completely satisfactory relativistic quantum-mechanical model*, … Therefore, I think that the *renormalization theory* was simply a way to sweep the difficulties of the divergences of electrodynamics under the rug".

Schwinger received the prize for his reformulation of the *relativistic* theory for the interaction between charged particles and electromagnetic fields as a consequence of the discovery that the electron's *magnetic moment* proved to be somewhat larger than expected. He solved this problem in 1948, when he was 30 years old, through "*renormalization*" and thereby contributed to a new quantum electrodynamics.

In his Nobel Prize lecture [Schwinger, J. (December 11, 1965). *Nobel Lecture. Relativistic Quantum Field Theory.*], Schwinger did not mention any of his 1948 and 1949 papers for which he was awarded the Nobel Prize. Instead, he focused on papers he had written between 1951 and 1965. He started by describing the logical foundations of *quantum field theory* or *relativistic quantum mechanics* which he defined *as the synthesis of quantum mechanics with relativity*. He noted improvements in the formal presentation of quantum mechanical principles by himself, described in a series of six papers on the theory of quantized fields published in *Physical Review* between June 1951 and June 1954, and by Feynman, described in Feynman's April 1948 *Space-Time Approach to Non-Relativistic Quantum Mechanics*, utilizing the concept of *action*, based on a study of Dirac concerning the correspondence between the quantum *transformation function* and the classical *action*. He identified two distinct formulations of quantum mechanics – his *differential formulation* utilizing a *differential* version of the *action* principle, and the *path integral formulation* of Feynman. He claimed that his *differential* version transcended the *correspondence principle* and incorporated on the same footing the two different kinds of quantum dynamical variable that are demanded empirically by the two known varieties of particles obeying Bose-Einstein or Fermi-Dirac statistics. The *quantum action principle was a differential statement about time transformation functions; all quantum-dynamical aspects of the system were derived from a single dynamical principle*. He also claimed that *quantum field theory* had failed no significant test nor could any decisive confrontation be anticipated in the near future, contrary to Feynman's reservations and Schwinger's statements in the preface to his 1958 book. He stated that *classical mechanics was a determinate theory*; knowledge of the *state* at a given time permits precise prediction of the result of measuring any property of the system. *Quantum mechanics was only statistically determinate*; it is the *probability* of attaining a particular result on measuring any property of the system not the outcome of an individual microscopic observation that is predictable from knowledge of the *state*. The *relativistic* structure of the *action principle* was completed by demanding that it present the same form independently of the particular partitioning of *space-time* into space and time. This was facilitated by the appearance of the *action operator* - the time integral of the Lagrangian - as the *space-time* integral of the Lagrange function. He also discussed a further eleven papers that he had published in *Physical Review* between July 1962 and October 1965 which attempted to extend his theory to gravitational fields and develop a *field theory of matter*.

"... not only in mechanics, but also in electrodynamics, no properties of observed facts correspond to a concept of absolute rest; but that for all coordinate systems for which the mechanical equations hold, the equivalent electrodynamical and optical equations hold also"

"...light is propagated in vacant space, with a velocity c which is independent of the nature of motion of the emitting body."

[Einstein, A. (1905). *Zur Elektrodynamik bewegter Körper*. (On the electrodynamics of moving bodies.) *Ann. Phys.*, 322, **10**, 891-921; translation by M. Saha (1920).]

This study concludes that, quite apart from enormity of the consequences of the two postulates of Albert Einstein's theory of special relativity, taken together, including *length contraction*, *time dilation*, and the requirement to assume a *point electron* in the unsuccessful attempt to introduce special relativity into quantum electrodynamics, the evidence in support of Einstein's *second postulate* on the constancy of the speed of light is far outweighed by the evidence against it. For this reason, until more satisfactory evidence in support of Einstein's *second postulate*, a refutation of the Ehrenfest paradox, and an explanation for the observed Doppler red shift and blue shift consistent with Einstein's two postulates, is provided, under any normal measure of a theory in physics, *Einstein's second postulate, and consequently his theory of special relativity, must be rejected.*

Part I provides annotated extracts and translations of the original papers leading up to the establishment of Einstein's *theory of special relativity* and Walter Ritz's competing *emission theory* of electromagnetic radiation. Part II examines the evidence in support of special relativity between 1905 and 2017. Part III addresses evidence against special relativity, reviews measurements of the speed of light, and provides the Conclusion.

The story in Part I starts with the experiment carried out by Hippolyte Fizeau in 1851 to measure the *relative speed of light in moving water*. [Fizeau, H. (1851). *Sur les hypothèses relatives à l'éther lumineux, et sur une expérience qui paraît démontrer que le mouvement des corps change la vitesse à laquelle la lumière se propage dans leur intérieur.* (The hypotheses relating to the luminous ether, and an experiment which appears to demonstrate that the motion of bodies alters the velocity with which light propagates itself in their interior.)] According to the theories prevailing at the time, light traveling through a moving medium would be dragged along by the medium, so that the measured speed of the light would be a simple sum of its speed through the medium plus the speed of the medium. Fizeau detected a dragging effect, but the magnitude of the effect that he observed was far lower than expected. When he repeated the experiment with air in place of water, he observed no effect. But the story of *special relativity* really originates with James Clerk Maxwell's publication of *A Dynamical Theory of the Electromagnetic Field* in 1865, in which Maxwell integrated Gauss's law, the Maxwell–Faraday version of Faraday's law of induction, Gauss's law for magnetism, and Ampère's law with Maxwell's addition for electricity and magnetism, to produce *Maxwell's equations for electromagnetism*. These demonstrated that electric and magnetic fields travel through space as waves moving at the speed of light in the same medium that is the cause of electric and magnetic phenomena. This medium in which light was presumed to travel was subsequently referred to as the *luminiferous ether*.

We then had the spectacle of grown men attempting to detect the *luminiferous ether*, most famously in the Michelson-Morley experiment of 1887 [Michelson, A. A. & Morley, E. W. (November, 1887). *On the relative motion of the Earth and the luminiferous ether.*] , in which an interferometer was used to measure the motion of the Earth relative to the ether by measuring it in perpendicular directions. The null result compelled Hendrik Lorentz, who in 1902 was awarded the Nobel Prize in Physics, and who was a leading advocate of the *principle of relativity*, to introduce his hypothesis of *length contraction* in 1892 [Lorentz, H.A. (1892). *La Théorie electromagnétique de Maxwell et son application aux corps mouvants.* (Maxwell's Electromagnetic Theory and its Application to Moving Bodies).], so that the velocity of light, c, remained constant, and the concept of the ether was preserved. This resulted in the *Lorentz transformation* in 1899 [Lorentz, H.A. (1899).*Simplified Theory of Electrical and Optical Phenomena in Moving Systems.*], under which, using Einstein's notation, lengths were contracted and times dilated by transformation equations $\xi = \beta(x - \upsilon t)$ and $\tau = \beta(t - \upsilon x/c^2)$, proportional to the *Lorentz factor* $\beta = 1/\sqrt{(1 - \upsilon^2/c^2)}$, when observed by an inertial observer moving in the direction of the x-axis with velocity υ. In 1906, Henri Poincaré published a paper [Poincaré, H. (1906). *Sur la dynamique de l'électron.* (On the Dynamics of the Electron.)] in which he noted that the Lorentz transformation was a merely a rotation in 4-dimensional space-time, described by $x^2 + y^2 + z^2 = c^2t^2$.

This is where Einstein enters the picture. Up until the publication of Einstein's seminal paper in September 1905 [Einstein, A. (1905). *Zur Elektrodynamik bewegter Körper.* (On the electrodynamics of moving bodies.)], the discussion of and experiments with light focused on the detection of the luminiferous ether in which electromagnetic waves were assumed to travel. In this paper, he introduced the two postulates of his theory of (special) relativity, in which "the "luminiferous ether" will prove to be superfluous". His *first postulate*, the *principle of relativity*, that *"... not only in mechanics, but also in electrodynamics, no properties of observed facts correspond to a concept of absolute rest; but that for all coordinate systems for which the mechanical equations hold, the equivalent electrodynamical and optical equations hold also ..."* was unexceptional, and had been the main driver behind the search for an alternative to current theories of electromagnetic waves based on an ether in a rest frame. In order to explain the Michelson-Morley result, he added his second postulate, that *"...light is propagated in vacant space, with a velocity c which is independent of the nature of motion of the emitting body"*, which was not. After stating the postulates, Einstein proceeded to work through the consequences of these assumptions on kinematics and electrodynamics. These consequences, which refer to observations by an *inertial observer*, were consequences of the *second postulate*, including the relativity of simultaneity, *length contraction* of a moving body, *time dilation* of the body in the moving system, the *relativistic velocity addition formula*, and the *relativistic* Doppler effect.

Einstein's *first postulate* states that the *same physical laws* which apply to observations of body in a stationary frame by an observer in a frame moving with constant velocity relative to the stationary frame *also apply to observations in the stationary frame of the body moving with the same velocity relative to the stationary frame*. Although this is only referred to obliquely by Einstein in 1911, it implies that *these consequences apply to any inertial body, without reference to an observer*, and, as such, result in *real consequences for moving bodies*. [Einstein, A. (1911). *Zum Ehrenfestschen Paradoxon. Eine Bemerkung zu V. Varîĉaks Aufsatz.* (On the Ehrenfest Paradox. A remark on V. Varîĉak's essay.)]

In order to provide a relativistic explanation for the Doppler red-shift was first noted by Fizeau, instead of simply applying time dilation, Einstein derived his *relativistic* Doppler effect by first assuming *non-relativistic* addition and subtraction of the velocity of the observer and the speed of light, as in the classical *non-relativistic* Doppler effect, *then adjusting this for time dilation*, [Einstein, A. (September, 1905), *Zur Elektrodynamik bewegter Körper.* § 7, Theory of Doppler's Principle and of Aberration.]. This was in contradiction of his postulates and of his assertion earlier in the same paper in § 5. The Composition of Velocities: "It follows, further, that *the velocity of light c cannot be altered by composition with a velocity less than that of light*". He also repeated this assertion and the derivation, but only in the form of a generalized Doppler principle, in his longer paper in *Jahrbuch der Radioaktivität.* [Einstein, A. (1907). *Relativitätsprinzip und die aus demselben gezogenen Folgerungen.* (On the Relativity Principle and the Conclusions Drawn from It.), §6. Application of the transformation equations to some problems in optics.

Despite this, the same *relativistic* Doppler effect calculation occurs in Feynman, R. P., Leighton, R. B. & Sands, M. (1963). *The Feynman Lectures on Physics.* Volume I. 34-7, § 34-6, and in modern text books.

In 1907, Einstein proposed a test for the relativity principle, based on Johannes Stark's measurements of the Doppler effect, by detecting and measuring the change in frequency of radiation emitted by moving canal rays, due to *time dilation* (subsequently referred to as the "*relativistic transverse Doppler effect*"). Einstein derived his formula for *time dilation*, which is of the second order in v/c, from the Lorentz transformation equation for time. This is also the correct calculation of the *relativistic* Doppler effect for longitudinal motion in accordance with Einstein's postulates, as indicated above. [Einstein, A. (1907). *Uber die Möglichkeit einer neuen Prüfung des Relativitätsprinzips.* (On the Possibility of a New Test of the Relativity Principle.)] This backfired; the numerical values were ten times greater than Einstein's formula.

Herman Minkowski entered the scene briefly with his 1908 lecture, published in 1909, [Minkowski, H. (1909). *Raum und Zeit.* (Space and Time.) in which, by restating Maxwell's equations as a symmetrical set of equations in the four variables (x, y, z, ict) combined with redefined vector variables for electromagnetic quantities, in what was subsequently known as *Minkowski spacetime*, he was able to show directly and very simply their invariance under Lorentz transformation. He died suddenly of appendicitis in Göttingen on January 12, 1909.

Then it was Walter Ritz's turn. From an early age Ritz exhibited a disposition for science and mathematics. Yet, also from an early age, his studies were hampered by recurring ill health. He first began to suffer from respiratory ailments at the age of nineteen, following a traumatic experience in September 1897, Climbing Mont Pleureur with friends, he looked back to see a group of them slip on fresh snow and plunge over a cliff; the emotional stress was compounded by physical overexertion and overexposure in the rescue efforts. Einstein, a year younger than Ritz, was a student and studied in the same section as Ritz at the Zurich Polytechnikum. The two registered for some courses with some of the same professors. Ritz made a better impression than Einstein at Zurich. By 1909, his reputation was such that a faculty committee at the University of Zurich considered Ritz to be the foremost of nine candidates to become their first professor of theoretical physics, noting that Ritz exhibited "an exceptional talent, bordering on genius". Ritz, however, had to be excluded from consideration because he was too ill to carry the workload, so the job went to Einstein instead.

In 1908 Ritz published a 130-page criticism of Maxwell–Lorentz electromagnetic theory, [Ritz, W. (1908). *Recherches critiques sur l'Électrodynamique Générale.* (Critical Research on General Electrodynamics.)]. *According to Ritz, the essential difficulties in electrodynamics were rooted in the field equations of electromagnetism.* He stressed that Maxwell's equations admitted far too many possible solutions, infinitely many in principle, and that this plethora of solutions involved absurd physical consequences. He argued that *advanced potentials* were devoid of physical significance; he denied the plausibility of convergent spherical waves; and he complained that Maxwell's equations allowed for the existence of a *perpetuum mobile. To avoid the ambiguous multiplicity of solutions of Maxwell's equations, Ritz claimed that retarded potentials had to be taken as fundamental.* These equations embodied a delay required for electromagnetic effects to traverse distances in space. By allowing only *retarded potentials*, only past states of a system could determine its present state, and *energy could be radiated only from matter*, rather than, say, be drawn out infinitely from a surrounding ether. As noted by Ehrenfest, [Ehrenfest, P. (1912). *Zur Frage nach der Entbehrlichkeit des Lichtäthers.* (On the question of the dispensability of the light ether.)] "The purpose of Ritz's study was to show that his theory of emissions provides an exact theory of relativity which, *in contrast to Einstein's theory of relativity, does not require any contraction of rigid bodies, no change in the rate of clocks*, and no assumption of the physical impossibility of propagation rates greater than $3 \cdot 10^{10}$ cm/sec".

Ritz died in Göttingen on July 7, 1909, at the age of 31, after seven weeks in the Göttingen medical clinic. After Ritz died the focus switched to the competing theories of electromagnetic radiation, but Einstein never referred to Ritz's theory, whilst vigorously promoting his own, and by 1925 Ritz's *emission theory* was effectively dead. In his last year and a half, Ritz had published a total of about four hundred pages of articles in the areas of theoretical spectroscopy, the foundations of electrodynamics, the problem of gravitation, and a method for the numerical solution of boundary-value problems. His collected works (541 pages), in German and French, were published two years later, in 1911. It appears that none of his papers were translated into English. The four papers included in this volume, including his joint paper with Einstein, were all translated into English by the author, using Microsoft Translator.

In 1909, Paul Ehrenfest published a paper [Ehrenfest, P. (1909). *Gleichförmige Rotation starrer Körper und Relativitätstheorie.* (Uniform Rotation of Rigid Bodies and the Theory of Relativity.)] on what was subsequently referred to as the *Ehrenfest Paradox*. In this paper Ehrenfest considers an ideally rigid cylinder that is made to rotate about its axis of symmetry. The radius as seen in the laboratory frame is always perpendicular to its motion and should therefore be equal to its value when stationary. However, the circumference should appear contracted to a smaller value than at rest, by the Lorentz factor, leading to a contradiction and which would cause the rotated Born rigid disk to shatter. In this paradox Ehrenfest correctly interprets Lorentz contraction according to Einstein's two postulates as being a *real contraction of a moving body*, not merely the contraction of a body as seen by an observer in an inertial reference frame.

Despite numerous attempts to resolve this for a rotating body between 1910 and 2007, they were not very convincing, and the paradox remained for a body moving in a straight line.

Part II examines the evidence which is normally used to support Einstein's *theory of special relativity*. It suffers from a number of problems. Some of it supports *emission theories* rather than *special relativity*, including Sagnac, G. (1913). *L'éther lumineux démontré par l'effet du vent relatif d'éther dans un interféromètre en rotation uniforme.*, Ives, H. E. & Stilwell, G. R. (1938). *An Experimental Study of the Rate of a Moving Atomic Clock,* Kantor, W. (1962). *Direct First-Order Experiment on the Propagation of Light from a Moving Source.,* and Suleiman, R. (2017). *The Sagnac Effect Falsifies Special Relativity Theory.*, of which Ives & Stilwell (1938) is the most notable.

Instead of demonstrating the existence of *time dilation* through the *relativistic* Doppler effect, which is of the second order in v/c, the red shift and blue shift and velocity data provided in the Ives & Stilwell (1938) paper provides *direct and precise confirmation of the classical non-relativistic (longitudinal) Doppler effect*, which is of the first order in v/c, and refutes the *relativistic* Doppler effect, due to time dilation, which only predicts a red shift (of the incorrect amount). *This, on its own, should have been enough to refute Einstein's second postulate.*

Some of the other evidence supports *both special relativity and emission theories*, including Michelson & Morley (1881, 1887), Tomaschek & Miller (1922, 1924), BonchBruevich (1956), Sadeh (1963), Alväger, Farley, Kjellman, & Wallin (1963), Filippas & Fox (1964), and Alväger, Farley, Kjellman, & Wallin (1964).

Some of it based on *celestial evidence* suffers from the consequences of the vast distance and the interstellar dust, etc, which lies between the source of the electromagnetic radiation and the observer, resulting in *extinction*. This includes Michelson & Morley (1881, 1887), Tolman (1910), de Sitter (1913), Tomaschek & Miller (1922, 1924), BonchBruevich (1956), and Brecher (1977).

There are also *extinction* problems with experiments with *light passing through a medium*, such as water or glass, which modify the velocity of light, including Kantor (1962), and Babcock & Bergman (1964).

There are also problems with observations on *rotating platforms* and other *accelerated* systems used to measure *time dilation* from the change in frequency (transverse Doppler effect) or the *Mössbauer effect*. This applies to Hay, Schiffer, Cranshaw & Egelstaff (1960), Kündig (1963), and Champeney, Isaak, & Khan (1965).

The evidence in support of the theory of special relativity reduces to extremely slim pickings. It includes the Frisch & Smith (1963) [Frisch, D. H. & Smith, J. H. (1963). *Measurement of the Relativistic Time Dilation Using μ-Mesons.*] *time dilation* experiment, based on the measurement of the number of μ-mesons incident at the top of Mt. Washington, New Hampshire, with speeds in the range between $0.9950\,c$ and $0.9954\,c$, which survived to reach sea level in Cambridge, Massachusetts; the Beckman & Mandics (1965) [Beckmann, P. & Mandics, P. (1965). *Test of the Constancy of the Velocity of Electromagnetic Radiation in High Vacuum.*] measurement of the *speed of light reflected from a moving mirror* in a high vacuum; Aleksandrov (1965) [Aleksandrov, E. B. (1965). *An Astrophysical Proof of the Second Postulate of Special Relativity Theory.*] observations on the *brightness variations of closest stars*; and, possibly, Brecher (1977) [Brecher, K. (1977). *Is the Speed of Light Independent of the Velocity of the Source?'s.*] observations of regularly pulsating X-ray sources in binary-stars. Of these, the Beckman & Mandics (1965) experiment seems most plausible, though may suffer from a potential problem with the use of a mirror.

PART III addresses evidence against Einstein's theory of special relativity, reviews measurements of the speed of light, and provides a Conclusion.

The primary evidence against Einstein's *theory of special relativity* comprises the Ehrenfest paradox, the *non-relativistic* longitudinal Doppler red shift and blue shift, and the failure of attempts to bring Einstein's theory of special relativity into mainstream quantum electrodynamics, including the requirement that the electron must be of zero radius and be treated as a point charge.

In the Conclusion it is noted that there is no evidence, based directly or indirectly, on the observation of the speed of electromagnetic radiation in a vacuum emitted by an inertial body or as observed by an inertial observer moving in a straight line and not involving mirrors. Quite apart from enormity of the consequences of Einstein's postulates taken together, including *length contraction*, *time dilation*, and the requirement to assume a *point electron*, the evidence in support of Einstein's *second postulate* is far outweighed by the evidence against it.

For this reason, until more satisfactory evidence in support of Einstein's second postulate, a refutation of the Ehrenfest paradox, and an explanation for the observed Doppler red shift and blue shift consistent with Einstein's two postulates, is provided, under any normal measure of a theory in physics, *Einstein's second postulate, and consequently his theory of special relativity, must be rejected.* Einstein's real contribution in his theory of special relativity was his first postulate, the *principle of relativity*, in helping to dispense with the luminiferous ether.

Einstein's Forward to Opticks, 1931:

Fortunate Newton, happy childhood of science! He who has time and tranquility can by reading this book live again the wonderful events which the great Newton experienced in his young days. Nature to him was an open book, whose letters he could read without effort. The conceptions which he used to reduce the material of existence to order seemed to flow spontaneously from experience itself, from the beautiful experiments which he ranged in order like playthings and describes with an affectionate wealth of detail. On one person he combined the experimenter, the theorist, the mechanic and, not least, the artist in exposition.

—Albert Einstein

This volume begins with a review of Sir Isaac Newton's laws of motion and of gravitation, his definitions of inertial and gravitational mass, and his equivalence principle, in his 1687 treatise on matter, *Philosophiæ Naturalis Principia Mathematica* (Mathematical Principles of Natural Philosophy), which provide the background to Einstein's theory. After struggling through Einstein's contortions, convoluted mathematics and obfuscations, one realizes what a joy this is. In contrast, in Einstein's own words, "unfortunately, I have immortalized my final errors in the academy-papers" … "it's convenient with that fellow Einstein, every year he retracts what he wrote the year before." Whilst Newton enumerated laws of nature based on observation and experiments, Einstein pursued theories, based on principles and assumptions, in search of evidence.

In *Principia*, Newton introduced his *laws of motion* and his *universal law of gravitation*. He also defined and distinguished between *inertial mass* and *gravitational mass* and showed them to be equal, and introduced his *equivalence principle* (between gravity and other forces causing acceleration).

In 1704, Newton published his treatise on light, *Opticks: or, a treatise of the reflexions, refractions, inflexions and colors of light. Also, two treatises of the species and magnitude of curvilinear figures.* Newton had written most of this in 1675, and read it at meetings of the Royal Society, but delayed publishing it. The Third Book was left "imperfect" so Newton proposed some queries so that these could be completed by others. These included two of the other issues co-opted by Einstein: "*Query* 1. Do not Bodies act upon Light at a distance, and by their action bend its Rays; and is not this action (*cæteris paribus*) strongest at the least distance?" and "*Quest.* 30. Are not gross Bodies and Light convertible into one another, and may not Bodies receive much of their Activity from the Particles of Light which enter their Composition?" He also noted that he had intended to make observations "on the making of the Fringes of Colors with the dark lines between them".

In a paper written in 1801 and published in 1804, Johann Soldner demonstrated, based on Newton's corpuscular theory of light, that a light ray passing a celestial body would be forced by the attraction of the body to describe a hyperbola whose concave side was directed against the attracting body, instead of progressing in a straight line. [Soldner, J. G. von. (1804). *Ueber die Ablenkung eines Lichtstrals von seiner geradlinigen Bewegung.* (On the Deflection of a Light Ray from its Rectilinear Motion by the attraction of a celestial body which passes nearby.)] He calculated the acceleration of gravity at the surface of the Sun and the angle of one leg of the hyperbola when a light ray was deflected by the Sun, giving $\omega = 0.84"$. He then doubled this number for a light ray passing the Sun, which describes two arms of the hyperbola, and obtained a deflection of 1.68", compared with the measured value 1.7", showing that the Newtonian theory was correct. Einstein obtained the same number in 1915, by importing Newton's law of gravitation.

In November 1905, Einstein published a follow-on from his September 1905 paper [Einstein, A. (September, 1905). *Zur Elektrodynamik bewegter Körper.* (On the Electrodynamics of Moving Bodies)]. This paper [Einstein, A. (November, 1905). *Ist die Trägheit eines Körpers von seinem Energieinhalt abhängig?* (Does the Inertia of a Body Depend Upon Its Energy Content?).] derived the conclusion that if a body gives off the energy L in the form of radiation, its mass diminishes by L/c^2, by applying the Lorentz transformation to the difference in kinetic energy of the light emitted by a body when referred to two systems of co-ordinates which are in motion relatively to each other, and assuming that the mass of a body is a measure of its energy-content. This formulation related only a change Δm in mass to a change L in energy without requiring the absolute relationship. By annotating the equivalent calculation according to *non-relativistic* Newtonian theory, we show that Newton's equations of motion result in $E = mc^2$ without the approximation that Einstein's calculation according to his *theory of special relativity* required.

Einstein's next paper [Einstein, A. (1907). *Über das Relativitätsprinzip und die aus demselben gezogenen Folgerungen.* (On the Relativity Principle and the Conclusions Drawn from It.)] begins with a summary of the development of Einstein's *theory of special relativity,* based on the difficulty of reconciling the negative result of Michelson and Morley's experiment with the existence of a luminiferous ether, and the failure to identify an *effect of the second order (proportional to v^2/c^2) predicted by* the Lorentz theory, which had required an ad hoc postulate according to which *moving bodies experience a certain contraction in the direction of their motion.* From this, instead of recognizing that Michelson and Morley's experiment was explained, with or without an ether, by his competitor and ex-classmate, Walter Ritz's, emission theory, Einstein concluded that

Lorentz's theory should be abandoned and replaced by a theory whose foundations correspond to the *principle of relativity*. Einstein assumed, based on the Michelson and Morley experiment, that physical laws are independent of the state of motion of the reference system, at least if the system is not accelerated; the *principle of relativity*. He also proposed that time should also be adjusted, so that the propagation velocity of light rays in a vacuum is everywhere equal to a universal constant c; the *principle of the constancy of the velocity of light*.

From this Einstein deduced that in order to maintain the same number of periods a clock in an inertial reference frame with velocity υ must complete $v = v_0 \sqrt{\{1 - \upsilon/c^2\}}$ periods per unit time, and move slower in the ratio $1 : \sqrt{\{1 - \upsilon/c^2\}}$ as observed from this system, than that of the same clock when at rest relative to that system. Einstein got *mass* into his theory using *Newton's second law of motion*, "force = mass x acceleration", applied to a charged electron, then expressing this in an inertial frame by applying a Lorentz transformation. This paper also marked the beginning of Einstein's long development of *general relativity*. He got *gravity* into his theory by adopting the "equivalence principle", the physical equivalence of a gravitational field and an acceleration of the reference system. Here, based on the *effect of the gravitational field on clocks*, in which a clock located in a gravitational potential Φ runs $(1 + \Phi/c^2)$ times faster, he derived the gravitational redshift, and the gravitational bending of light. Einstein returned to these topics in 1911.

In 1911 [Einstein, A. (1911). *Über den Einfluss der Schwerkraft auf die Ausbreitung des Lichtes*. (On the Influence of Gravitation on the Propagation of Light.)] Einstein returned to the question that he tried to answer in his 1907 paper, *whether the propagation of light is influenced by gravitation*. He began with the hypothesis that the physical nature of the gravitational field is based on the "*equivalence principle*", that matter subject to *uniform acceleration* is physically equivalent to matter in a *gravitational field*. He noted that the *theory of relativity* showed that the *inertial mass* of a body increased with the *energy* it contained and extended this to *gravitational mass*. Einstein also observed that *radiation* emitted in a *uniformly accelerated system* from one point, when the velocity of the reference frame was υ, would arrive at another point at a distance h when the time h/c has elapsed and the velocity is γh/c (where γ is the acceleration). From this, Einstein deduced that according to the *theory of special relativity* the energy of the *radiation* arriving at the second point has increased by $E_2 = E_1 (1 + \upsilon/c) = E_1 (1 + \gamma h/c^2)$. Then according to the "*equivalence principle*", $E_2 = E_1 + E_1/c^2 \Phi$, where Φ is the difference in *gravitation potential* between the two points.

Einstein made an incorrect deduction that substitution of γh/c for the velocity υ in the classic non-relativistic Doppler effect $v' = v(1 \pm \upsilon/c)$ made the *change in frequency* of light a function of the *acceleration* γ, and consequently of the *gravitational potential*

$\Phi = \gamma h$. From this he claimed that the *frequency* of the light will also have increased in a *uniformly accelerated system* $v_2 = v_1 (1 + \gamma h/c^2)$ according to the classic non-relativistic Doppler effect (*which assumes subtraction and addition of velocities*), and consequently by $v_2 = v_1 (1 \pm \Phi/c^2)$ in a *gravitational field*. Based on this, Einstein found that *the spectral lines of sunlight, as compared with the corresponding spectral lines of terrestrial light sources, must be somewhat displaced toward the red*, by the relative amount $(v_0 - v)/v_0 = -\Phi/c^2 = 2 \times 10^{-6}$, where Φ is the (negative) difference between the *gravitational potential* between the surface of the Sun and the Earth. Einstein addressed the apparent difference in the number of periods per second of the light at the two locations with different *gravitational potentials* by defining *time* at the two locations so that the number of wave crests and troughs was the same, but *this resulted in the speed of light in a gravitational field no longer being constant but a function of the location,* given by the relation, $c = c_0 (1 + \Phi/c^2)$. From this he calculated that a light-ray going past the Sun would undergo deflection by $4 \times 10^6 = 0.83$ *seconds of arc*. This number was subsequently doubled by Einstein in 1915, by substituting Newton' law of gravitation for his equation based on Euler's equation, which brought it in line with the Newtonian result first calculated by Soldner in 1801.

In 1911, Max Abraham formulated an alternative theory of gravitation based on *static gravitational fields* in terms of Hermann Minkowski's four-dimensional space-time formalism and Einstein's 1911 relation between the variable velocity of light and the gravitational potential. [Abraham, M. (1911). *Zur Theorie der Gravitation.* (On the New theory of Gravitation.) and Abraham, M. (1912). *Berichtigung.* (Correction.)]. Abraham introduced the general four-dimensional line element involving a variable metric tensor. However, for the time being Abraham's expression remained an isolated mathematical formula without context and physical meaning which, at this point, was neither provided by Abraham's nor by Einstein's physical understanding of gravitation.

Einstein, A. (1911) [*Über den Einfluss der Schwerkraft auf die Ausbreitung des Lichtes.* (On the Influence of Gravitation on the Propagation of Light.)] had shown that the validity of one of the fundamental laws of that theory, namely, the law of the *constancy of the speed of light*, could claim to be valid only for space-time domains of constant gravitational potential. In February, 1912, [Einstein, A. (February, 1912). *Lichtgeschwindigkeit und Statik des Gravitionsfeldes.* (The Speed of Light and the Statics of the Gravitational Field.)] Einstein noted that despite the fact that this result *excluded the general applicability of the Lorentz transformation*, it should not deter us from pursuing the consequences of that path. Here he took that further by demonstrating that the Lorentz transformation could not be established for infinitely-small space-time regions either *as soon as one abandons the universal constancy of c.*

In Abraham, M. (1912). *Relativität und Gravitation. Erwiderung auf eine Bemerkung des Hrn. A. Einstein*, Abraham criticized Einstein's way of arriving at his results. He did not like Einstein's use of the "equivalence hypothesis", and the correspondence between reference systems. It appeared to Abraham as a fluctuating basis, because Einstein had not yet adopted the space-time formalism of relativity. In July 1912 [Einstein, A. (July, 1912). *Relativität und Gravitation: Erwiderung auf eine Bemerkung von M. Abraham.* (Relativity and Gravitation. Reply to a Comment by M. Abraham.)], Einstein responded to Abraham's criticism that by abandoning the postulate of the *constancy of the speed of light* and by renouncing the invariance of the systems of equations in relation to Lorentz transformations, he had sacrificed the *theory of relativity*. Einstein argued that the fact that the *principle of the constancy of the velocity of light* could be maintained only insofar as one restricts oneself to spatio-temporal regions of *constant gravitational potential* was not the limit of validity of the *principle of relativity*, but that of the *constancy of the velocity of light*, and thus of the current theory of relativity.

In October 1912 [Abraham, M. (1912). *Una nuova theoria della gravitazione.* (A new theory of gravitation.)], Abraham elaborated on his *new theory of gravitation*. He began by making an analogy between gravitation and electromagnetism from which he concluded that although the strict analogy must be renounced the essential viewpoints of Maxwell's theory must be retained, namely that *the fundamental laws must be differential equations that describe the excitation and propagation of the gravitational field*, and a positive energy density and an energy current must be assigned to that field. He proposed *a new theory of the gravitational field* based on the hypothesis in Einstein, A. (1911) [*Über den Einfluss der Schwerkraft auf die Ausbreitung des Lichtes.* (On the Influence of Gravitation on the Propagation of Light.)] that *the speed of light depended upon the gravitational potential*. He began with a Lagrangian function $L = - mc^2 f(v/c)$ that was valid for the dynamics of electrons, in which v signified the velocity, and m was the *inertial rest mass*. From this he obtained the values of the *impulse* and *energy* from $G = \partial L/\partial v$, and $E = v \partial L/\partial v - L$, and the *equations of motion* $d/dt (\partial L/\partial x') - (\partial L/\partial x) = 0$, etc. In the case of constant c, the Lagrangian function would depend upon only the velocity v, but not position, so only the first terms that would enter into the *equations of motion* were the ones that contained the derivatives of the components of *impulse* with respect to time $d/dt (\partial L/\partial v \cdot x'/v) = d/dt (G x'/v) = d\mathbf{G}_x/dt$, etc.

In the new theory of c, the Lagrangian *also depended upon the coordinates*, so the second terms in the Lagrange equations, $\partial L/\partial x = \partial L/\partial c \cdot \partial c/\partial x$, etc. need to be retained. These represented the components of a force that is proportional to the *gradient* of c, and which, according to his first postulate, was *the force of gravity*. The *equations of motion* could then be written in vectorial form $d\mathbf{G}_x/dt = \partial L/\partial c$ grad c. These were exact for the free motion of a material point in the gravitational field, but they also applied to a system whose dimensions were small enough that it could be *equated to a material point.*

Abraham applied his *third postulate that made gravitation proportional to the energy of a moving point,* $\partial L/\partial c = -\chi$. E, so the *gravitational force* became K = $-\chi(c)$. E . grad c, and from $\chi(c) = 1/c$, K = $-$ E/c . grad c; and from E = M . c, the force acting on a point at rest became K = $-$ M . grad c, where M = cm. The *force that acted on a material point in a given gravitational field* was then determined by assuming that the mass was proportional to the energy regarded as the source of the gravitational field. Setting u = $\sqrt{c}$ and $\square$u = Δu $-$ 1/c ∂u/∂t (1/c ∂u/∂t), where $\square$ denotes the operator $\sum_\tau \partial^2/\partial x_\tau^2$ (τ = 1 to 4), this led to the *fundamental equation of the gravitational field,* $\square$u = 2α. η/u for *matter in motion,* where η was the *energy density of the matter,* and α a universal constant, *which coupled the attracting mass of a body to its energy.* For the *static field,* this became Δu $\equiv$ div grad u = 2α. η/u = 2α μu, where μ was the "specific density" of matter, so that the divergence of grad u was proportional to the *density of matter.* This could also be written in the form Δc = div grad c = 4α (η + ε), where ε was the *energy density of the gravitational field,* i.e. the divergence of the gradient of c was proportional to the total energy density in the static field. Applied to a homogeneous sphere, the *radial gradient of c* became dc/dr = 2c_0 ϑ/r^2 (1 $-$ ϑ/r), and since gravity was proportional to this gradient, *Newton's law was not exact according to his theory* by a factor $-$ E_e/E_t . a/r = 10^{-6} in the case of the Sun. Abraham went on to show that this difference was due to the *energy of the gravitational field outside the sphere.* He also showed that his theory of gravitation contradicted the second postulate of the *theory of special relativity,* even infinitesimally.

In Einstein, A. & Grossmann. M. (1913). *Entwurf einer verallgemeinerten Relativitätstheorie und eine Theorie der Gravitation.* (Outline of a generalized theory of relativity and a theory of gravitation.), Einstein made a new attempt, in collaboration with mathematician Marcel Grossmann, at a *relativistic theory of gravitation,* known as the "Entwurf (outline) theory". He went back to basics, expressing the scalar *gravitational field* in terms of a symmetric, four-dimensional metric tensor, as a generalization of the Poisson equation of *Newton's law of gravitation.* Based on his *"equivalence principle"* (of gravitation and a uniformly accelerated reference frame), from the equation δ {$\int$ds} = 0, he derived the *equations of motion* d/dt {mx'/$\sqrt{(c^2 - q^2)}$} = $-$ (mc δc/δx) /$\sqrt{(c^2 - q^2)}$, where q was the *translational velocity* of the system, in which the right side of these equations *represented the force* $\mathfrak{R}_x$ *exerted by the gravitational field on the mass point.* For the special case of rest, where q = 0, in a static gravitational field $\mathfrak{R}_x = -$ m δc/δx; *from which Einstein concluded that c played the role of gravitational potential.*

In order to uphold the *principle of relativity,* Einstein generalized the theory of relativity in such a way that it contained the theory of the *static gravitational field* as a special case. He substituted $ds'^2 = g_{11}dx'^2 + g_{22}dy'^2 + \ldots + 2g_{12}dx' \, dy' + \ldots = \Sigma_{\mu\nu} \, g_{\mu\nu}dx_\mu dx_\nu$, for $(- dx^2 - dy^2 - dz^2 + c^2dt^2)$ to produce an equation of the form $\delta \, \{\int ds'\} = 0$, where the quantities $g_{\mu\nu}$ were functions of x', y', z', t', so that, in the general case, the *gravitational field* was characterized by ten space-time functions. From the meaning that ds played in the *law of motion* of the material point Einstein concluded that ds must be an *absolute invariant* (scalar), and that the quantities $g_{\mu\nu}$ formed a covariant tensor of the second rank, which he called the *covariant fundamental tensor*.

Einstein attempted to obtain the differential equations that determined the quantities g_{ik}, i.e. the *gravitational field*, from a generalization of Poisson's equation $\Delta\varphi = 4\pi k\rho$, where φ was the *gravitational potential*, k the gravitational constant, and ρ the mass density. As he was unable to find a direct solution, he introduced some assumptions whose correctness "seemed plausible, but was not evident". He assumed a generalization of the form
$\kappa. \, \Theta_{\mu\nu} = \Gamma_{\mu\nu}$, where the tensor $\Theta_{\mu\nu}$ was the (contravariant) voltage-energy tensor of the material flow, κ a constant, and $\Gamma_{\mu\nu}$ a contravariant tensor of the second order, which emerged from the fundamental tensor $g_{\mu\nu}$ by differential operations. He then introduced the following abbreviations where $\gamma_{\mu\nu}$ was the fundamental tensor:
$- 2\kappa. \, \vartheta_{\mu\nu} = \Sigma_{\alpha\beta\tau\rho} \, (\gamma_{\alpha\mu} \, \gamma_{\beta\nu} \, \partial g_{\tau\rho}/\partial x_\alpha \, \partial\gamma_{\tau\rho}/\partial x_\beta - \frac{1}{2} \, \gamma_{\mu\nu} \, \gamma_{\alpha\beta} \, \partial g_{\tau\rho}/\partial x_\alpha \, \partial\gamma_{\tau\rho}/\partial x_\beta)$, in which he designated $\vartheta_{\mu\nu}$ as the "*contravariant stress-energy tensor of the gravitational field*"; and
$- 2\kappa. \, t_{\mu\nu} = \Sigma_{\alpha\beta\tau\rho} \, (\partial g_{\tau\rho}/\partial x_\mu \, \partial\gamma_{\tau\rho}/\partial x_\nu - \frac{1}{2} \, g_{\mu\nu} \, \gamma_{\alpha\beta} \, \partial g_{\tau\rho}/\partial x_\alpha \, \partial\gamma_{\tau\rho}/\partial x_\beta)$, where the covariant tensor reciprocal to it was denoted by $t_{\mu\nu}$. From this he obtained the *gravitational equations*
$\kappa. \, \Theta_{\mu\nu} = \Gamma_{\mu\nu}$, in the form $\Delta_{\mu\nu} \, (\gamma) = \kappa \, (\Theta_{\mu\nu} + \vartheta_{\mu\nu})$ and $- D_{\mu\nu} \, (\gamma) = \kappa \, (t_{\mu\nu} + T_{\mu\nu})$, in terms of the sum of the *stress-energy tensors of the gravitational field and matter*. He still had to introduce a link to the weak attractive gravitational force.

The *Einstein–Besso Manuscript*, which was sold at auction in Paris on November 23, 2021, for a record $11.5 million, comprised a stack of about 50 pages of scratchpad of attempts by Einstein and his close friend Michele Besso from the spring of 1913 to calculate the precession of the perihelion of Mercury according to Einstein's new theory of relativity. The end result given in the manuscript was 1821″ or 30′ of arc (Figure 1), more than three times the *total motion* of Mercury's perihelion. There are indications in the manuscript that Besso found the trivial error that led them to overestimate the effect by a factor of 100. Yet 18 was still 25 shy of 43. They tried to make the theory yield a few more seconds in other ways but came up empty. In early 1914, Einstein mailed what they got so far to Besso and urged his friend to keep working on the project. Besso tried but made no further progress. On Page 61 of Einstein's notebook (1909-14) the correct numbers are inserted in an expression for the perihelion advance of Mercury, which is a good approximation of the expression given in the Einstein-Besso manuscript, and the end result is given as 17″. Einstein had been trying since 1907 but it was not until November 1915 that he came up with a calculation that yielded 43″.

In 1913, Gunnar Nordström published his *theory of gravitation* which was a predecessor of general relativity [Nordström, G. (1913). *Träge und schwere Masse in der Relativitätsmechanik.* (Inertial and gravitational mass in relativity mechanics.)]. His theory was *the first known example* of a *metric theory of gravitation*, in which the effects of gravitation are treated entirely in terms of the geometry of a curved spacetime. From the *proportionality of inertial and gravitational mass*, Nordstrom deduced that the field equation should be $\varphi \Box \varphi = -4\pi\rho\, T_{matter}$, which is nonlinear, and the *equation of motion* to be $d(\varphi\, u_a)/ds = -\varphi_{,a}$ or $\varphi\, u_a = -\varphi_{,a} - \varphi\cdot u_a$.

Uncharacteristically, Einstein took the first opportunity to proclaim his approval of the new theory in a keynote address to the annual meeting of the Gesellschaft Deutscher Naturforscher und Arzte (Society of German Scientists and Physicians), on September 23, 1913. Einstein showed that the contribution of *matter* to the *stress–energy tensor* should be $(T_{matter})_{ab} = \varphi\, \rho\, u_a\, u_b$, and derived an expression for the *stress–energy tensor* of the *gravitational field* in Nordström's theory to be $4\pi\, (T_{grav})_{ab} = \varphi_{,a}\, \varphi_{,b} - \frac{1}{2}\, \eta_{ab}\, \varphi_{,m}\, \varphi^{,m}$, which he proposed should hold in general, and showed that *the sum of the contributions to the stress–energy tensor from the gravitational field energy and from matter* would be *conserved*. He also showed that the *field equation* of Nordström theory followed from the Lagrangian $L = 1/8\pi\, \eta^{ab}\, \varphi_{,a}\, \varphi_{,b} - \rho\varphi$, and that his theory could be derived from an *action principle*.

In his lecture to the 85th meeting of the Gesellschaft Deutscher Naturforscher und Arzte in Vienna on September 23, 1913 [Einstein, A. (1913). *Zum gegenwärtigen Stande des Gravitationsproblems.* (On the present state of the problem of gravitation.)], Einstein began with the observation that Newton's *action-at-a-distance* gravitational theory needed to be extended in order to comply with relativity theory, under which it is impossible to send signals with a velocity greater than that of light. He proposed four postulates, which could be employed by a *gravitational theory*. He then presented two generalizations of Newton's theory which he claimed were, in the present state of our knowledge, the *most natural*.

First, he introduced *Nordström's theory of gravity*. He noted that according to the theory of relativity together with the theory of gravitation, an isolated material point moved uniformly in a straight line according to Hamilton's equation $\delta (\int d\tau) = 0$, where $d\tau = \{\surd(c^2 dt^2 - dx^2 - dy^2 - dz^2) = dt\, \surd(c^2 - q^2)$; or $\delta (\int H\, dt) = 0$; $H = -m\, d\tau/dt = -m\, \surd(c^2 - q^2)$ was the Lagrangian of the moving point; and m is a constant characteristic of it, its "*mass*". *In Nordström's theory* this was obtained *by assuming that the covariance of the equation with respect to linear orthogonal substitutions still stood.* According to this theory the *gravitational field* could be described by a scalar, and the motion of the material point in a gravitational field could be described by an equation of Hamiltonian form; and light rays were not bent by the gravitational field.

In Einstein's theory, the general equation for the *gravitational field*, viewed as a generalization of Poisson's equation for the *gravitational field*, was obtained by setting $-\kappa \sum T_{\sigma\sigma} = \varphi \square \varphi$, where κ denoted a universal constant (gravitational constant), and $\square$ denoted the operator $\sum_\tau \partial^2/\partial x_\tau^2$ ($\tau = 1$ to 4), resulting in $\sum_v \partial/\partial x_v (T_{\mu v} + t_{\mu v}) = 0$, where $T_{\mu v}$ was the *stress-energy tensor of matter* and $t_{\mu v}$ the component of the *stress-energy tensor of the gravitational field*. Einstein's *equations for the gravitational field*, based on a generalization of Poisson's equation, in which the *gravitational field* was determined by the ten quantities $g_{\mu v}$ instead of by φ, and the ten-component symmetric tensor $\Theta_{\mu v}$ was the field source in place of ρ, were then introduced resulting in equations of the form $\Gamma_{\mu v} = \kappa \Theta_{\mu v}$, where $\Gamma_{\mu v}$ was a differential expression formed from the quantities $g_{\mu v}$. From this Einstein developed the desired *equations of the gravitational field*, the *momentum-energy equation for the material process and gravitational field together*, and the *conservation law* in the form $\sum \partial/\partial x_v (\boldsymbol{T}_{\sigma v} + \boldsymbol{t}_{\sigma v})$, where $\boldsymbol{T}_{\sigma v}$ and $\boldsymbol{t}_{\sigma v}$ were the *stress components of matter and the gravitational field*. Einstein obtained Newton's system through a series of approximations resulting in $g^*_{44} = \kappa c^2/4\pi \int \rho_0\, dv/r$, where the integration was extended over *three-dimensional space*, and r denoted the distance between dv and the source. He then asserted that "*the customary gravitational constant K is connected here with our constant κ by the relation $K = \kappa c^2/8\pi$*", from which it followed that $K = 6.7.10^{-8}$ and $\kappa = 1.88.10^{-27}$ cm g^{-1}. *This relation effectively substituted the Newtonian equation for Einstein's relativistic equation* in which the link to matter was based on Euler's equation, and in which the gravitational force was far too large and of opposite sign.

The discussion following Einstein's lecture revealed that no-one present appeared to have heard of, or did not believe, Soldner's 1801 calculation of the bending of light based on Newton's law of gravitation and the corpuscular theory of light.

In a response [Einstein, A. (1914). *Principielles zur verallgemeinerten Relativitätstheorie und Gravitationstheorie.* (On the foundations of the generalized theory of relativity and the theory of gravitation.)] to a question by Gustav Mie on the relationship between the Einstein and Grossmann paper [(1913). *Entwurf einer verallgemeinerten Relativitätstheorie und eine Theorie der Gravitation.*] and Hermann Minkowski's work, Einstein noted that Minkowski founded a four-dimensional covariant theory on the invariant $ds^2 = \sum dx_v^2$ which provided the equations of the original *theory of relativity*. In an analogous way, a covariant theory could be based on the invariant $ds^2 = \sum_{\mu v} g_{\mu v} dx_\mu dx_v$, by means of the "*absolute differential calculus*", which provided the corresponding equations of the *new theory of relativity*. Einstein also provided a summary of the Einstein/Grossman generalized theory of gravitation leading to the differential equations $\sum_{\alpha\beta\mu} \partial/\partial x_\alpha \{\sqrt{(-g)}\, \gamma_{\alpha\beta}\, g_{\sigma\mu}\, \partial\gamma_{\mu v}/\partial x_\beta\} = \kappa(\boldsymbol{T}_{\sigma v} + \boldsymbol{t}_{\sigma v})$, where $-2\kappa\, \boldsymbol{t}_{\sigma v} = \sqrt{(-g)}\{\sum_{\beta\tau\rho} \gamma_{\beta v}\, \partial g_{\tau\rho}/\partial x_\sigma\, \partial\gamma_{\tau\rho}/\partial x_\beta\} - \frac{1}{2} \sum_{\alpha\beta\tau\rho} \delta_{\sigma v}\, \gamma_{\alpha\beta}\, \partial g_{\tau\rho}/\partial x_\alpha\, \partial\gamma_{\tau\rho}/\partial x_\beta\}$, after arbitrarily adding the factor -2κ, where $\delta_{\sigma v} = 1$ or 0, depending on $\sigma = v$ or $\sigma \neq v$. He clarified that c was not to be understood as a constant, but as a function of space coordinates ($c = \sqrt{g_{44}}$), which was a measure of the gravitational potential.

In February 1914, Einstein and a student of Lorentz', Adriaan Fokker, who visited Einstein in Zurich, [Einstein, A. & Fokker, A. (1914). *Die Nordströmsche Gravitationstheorie vom Standpunkt des absoluten Differentialkalküls.* (Nordström's theory of gravitation from the point of view of the absolute differential calculus.)] showed that it was possible to arrive at a complete representation of *Nordström's theory of gravitation* by using the invariant *absolute differential calculus*, by first referring the *four-dimensional manifold* to totally arbitrary coordinates (corresponding to Gaussian coordinates in the theory of surfaces), and only restricting the choice of the reference system when required. It turned out that one arrived at Nordström's theory, rather than at the Einstein-Grossmann theory, if one made *the single assumption that it was possible to choose privileged reference systems in such a way that the principle of the constancy of the velocity of light would be preserved.*

The difference between the two theories was that according to the Einstein-Grossmann theory, the *gravitational field* is determined by ten quantities $g_{\mu\nu}$, for which ten formally equivalent equations were given, whilst *Nordström's theory* amounted to the assumption that with an appropriate choice of the reference system the ten quantities $g_{\mu\nu}$ could be reduced to a single quantity Φ^2. Einstein showed that *in order to determine Φ^2, a single differential equation was required that had a scalar character like Poisson's equation*, which was completely determined by *the assumption that it was of the second order* if one also took into account the fact that *it must be a generalization of Poisson's equation* of the form $\Gamma = \kappa \, \mathbf{T}$, where Γ was a scalar formed from the quantities $g_{\mu\nu}$ and their first and second derivatives, and $\mathbf{T}$ was a scalar determined by the material process, by the $\mathbf{T}_{\sigma\nu}$, and κ denoted a constant. Then by selecting a reference system with respect to which the principle of the constancy of the velocity of light was satisfied the components $g_{\mu\nu}$ of the *fundamental tensor* were reduced, to $\Sigma_{\mu\nu} \, g_{\mu\nu}dx_\mu dx_\nu = \Phi^2 dx_1^2 + \Phi^2 dx_2^2 + \Phi^2 dx_3^2 - \Phi^2 dx_4^2$ where $x_1 = x$, $x_2 = y$, $x_3 = z$ and $x_4 = ct$. The *momentum and energy equations for matter* took the form $\Sigma \, \partial T_\nu/\partial x_\nu = \partial \log\Phi/\partial x_\sigma \, \Sigma \, \mathbf{T}_{\tau\tau}$, according to which *only the scalar* $\{1/\sqrt{(-g)}\} \, \Sigma_\tau \, \mathbf{T}_{\tau\tau}$ *determined the influence of the gravitational field on a system.* The differential equation of the gravitational field took the form $1/\Phi^3 \, (\partial^2\Phi/\partial x_1^2 + \partial^2\Phi/\partial x_2^2 + \partial^2\Phi/\partial x_3^2 - \partial^2\Phi/\partial x_4^2) = k/\Phi^4 \, \Sigma_\tau \, \mathbf{T}_{\tau\tau}$, where k denoted a new constant, or $\Phi \, \square \, \Phi = k \, \Sigma_\tau \, \mathbf{T}_{\tau\tau}$.

In November 1914, Einstein published a paper [Einstein, A. (November, 1914). *Die formale Grundlage der allgemeinen Relativitätstheorie.* (The formal foundations of the general theory of relativity.)] of which the primarily objective was to provide a formal mathematical treatment of the *metric tensor theory* introduced in his previous papers. Section A provided the basic ideas of his theory, in particular the replacement of $ds^2 = \Sigma_\nu \, dx_\nu^2$ by $ds^2 = \Sigma_{\mu\nu} \, g_{\mu\nu} \, dx_\mu \, dx_\nu$. Section B provided simple deductions for the basic laws of absolute differential calculus to enable the reader to grasp the theory completely without reading other purely mathematical treatises. This particularly related to four-vectors, covariant tensors of second and higher ranks, including the covariant fundamental tensor $g_{\mu\nu}$, and the formation of tensors by differentiation. It also introduces the equation governing the movement of a material point in a gravitational field $\delta\{\int ds\} = 0$, which corresponds to a geodesic line in a four-dimensional manifold. In Section C, he derived the Eulerian equations of hydrodynamics and the field equations of the electrodynamics of moving bodies, in order to illustrate the mathematical methods.

In Section D, Einstein derived his *field equations* based on the assumption that covariant V-tensor $\mathbf{G}_{\mu\nu} = \partial H\sqrt{(-g)}/\partial x^{\mu\nu} - \Sigma_\sigma\, \partial/\partial x_\sigma\, (\partial H\sqrt{(-g)}/\partial g_\sigma^{\mu\nu})$ had a fundamental role in the field equations of gravitation, and that those equations had to take the place that Poisson's equation had in the Newtonian theory. He assumed that these equations would have a strong correlation between the tensors $\mathbf{G}_{\mu\nu}$ and $\mathbf{T}_\sigma^\nu$ because *the energy tensor T_σ^ν was decisive for the action of the gravitational field upon matter*, so assumed that his *field equations* were of the form $\mathbf{G}_{\sigma\tau} = \kappa\, \mathbf{T}_{\sigma\tau}$ where κ was a universal constant and $\mathbf{T}_{\sigma\tau} = \Sigma_\nu\, g_{\nu\tau}\, \mathbf{T}_\sigma^\nu$ was the symmetric covariant V-tensor, associated with the mixed *energy tensor* $\mathbf{T}_\sigma^\nu = \Sigma_\tau\, \mathbf{T}_{\sigma\tau}\, g^{\nu\tau}$. The ten equations $\mathbf{G}_{\sigma\tau} = \kappa\, \mathbf{T}_{\sigma\tau}$ could then be used to determine the ten functions $g^{\mu\nu}$ if the $\mathbf{T}_{\sigma\tau}$ were given. Without using any physical knowledge of gravitation, Einstein arrived at his *differential equations of the gravitational field* in a purely covariant-theoretical manner. Setting $H = \frac{1}{4}\, \Sigma_{\alpha\beta\tau\rho}\, g^{\alpha\beta}\, \partial g_{\tau\rho}/\partial x_\alpha\, \partial g^{\tau\rho}/\partial x_\beta$, he obtained the formulations $\Sigma_{\alpha\beta}\, \partial/\partial x^\alpha\, \{\sqrt{(-g)}\, g^{\alpha\beta}\, \Gamma_{\alpha\beta}^\nu/\partial x_\beta\} = -\kappa(\mathbf{T}_\sigma^\nu + \mathbf{t}_\sigma^\nu)$, where $\Gamma_{\alpha\beta}^\nu = \frac{1}{2}\, \Sigma_\nu\, g^{\nu\tau}\, \partial g_{\sigma\tau}/\partial x_\beta$, and $\mathbf{t}_\sigma^\nu = \sqrt{(-g)}/\kappa\, \{g^{\nu\tau}\, \Gamma_{\mu\sigma}^\rho\, \Gamma_{\rho\tau}^\mu - \frac{1}{2}\, \delta_\sigma^\nu\, g^{\tau\tau'}\, \Gamma_{\mu\sigma}^\rho\, \Gamma_{\rho\tau'}^\mu\}$.

In section E, Einstein showed that Newton's law of gravitation resulted from his *theory of general relativity* as an approximation. Einstein failed to provide any representation of the *weak attractive gravitation force between matter*, so the constant κ that was introduced into his differential equations of the gravitational field in terms of the energy tensors was of the wrong sign and far too large. *In order to calculate* the redshift of light, and bending of light in a gravitational field*, he substituted the gravitational potential from Newton's law of gravitation*. The resulting calculations were consequently the Newtonian results.

In the first of three papers [Einstein, A. (November 4, 1915). *Zur allgemeinen Relativitätstheorie.* (On the General Theory of Relativity.)] published by Einstein in November 1915 that led to the final *field equations* for *general relativity*, Einstein recognized that what he had thought in his November 1914 paper [*Die formale Grundlage der allgemeinen Relativitätstheorie.*] was the only *law of gravitation* which corresponded to the general postulates of relativity could not be proved at all on the path taken there. He had assumed that the postulate of relativity was always fulfilled if one takes the Hamilton principle as a basis; but in reality, *it did not provide a means of determining the Hamilton function H of the gravitational field*. The equation $S_\sigma^\nu = 0$, which restricted the choice of

H, expressed nothing other than that H is supposed to be an invariant with respect to linear transformations, *which requirement had nothing to do with that of the relativity of acceleration. The "Entwurf" field equations were untenable.*

His new theory rested on the postulate of the *covariance of all systems of equations relative to transformations with the substitution determinant 1*; the equation valid for arbitrary substitutions $d\tau' = \partial(x_1'\ldots x_4')/\partial(x_1\ldots x_4)\, d\tau$, due to the premise in the new theory $\partial(x_1'\ldots x_4')/\partial(x_1\ldots x_4) = 1$, became $d\tau' = d\tau$, so that the four-dimensional volume element was an invariant. The new *field equations* became $\sum_\alpha \partial\Gamma^\alpha_{\mu\nu}/\partial x^\alpha + \sum_{\alpha\beta} \Gamma^\alpha_{\mu\beta}\, \Gamma^\beta_{\nu\alpha} = -\,\kappa\, T_{\mu\nu}$.

In an addendum to the previous paper, [Einstein, A. (November 11, 1915). *Zur allgemeinen Relativitätstheorie.* (On the General Theory of Relativity). (Addendum.)], Einstein introduced a new approach in which he assumed, by analogy with the vanishing of the scalar $\sum_\mu T_\mu{}^\mu$ for the electromagnetic field, that the energy tensor $T_\mu{}^\lambda$ of "matter", to which the previous expression related, might vanish. He suggested that *it might very well be that in "matter" gravitational fields formed an important constituent.* The only difference in content between the field equations derived from *general covariance* and those of the recent paper was that the value of $\sqrt{(-g)}$ could not be prescribed in the latter. This value was rather determined by the equation $\sum_{\alpha\beta} \partial/\partial x_\alpha \{g^{\alpha\beta}\, \partial lg\sqrt{(-g)}/\partial x_\beta\} = -\,\kappa \sum_\sigma T_\sigma{}^\sigma$. In this paper, Einstein showed that this equation implied $\sqrt{(-g)}$ could only be constant *if the scalar of the energy tensor vanished.* Under the new derivation $\sqrt{(-g)} = 1$. The *vanishing of the scalar of the energy tensor of "matter"* then followed from the *field equations* instead of from the equation $\sum_{\alpha\beta} \partial/\partial x_\alpha \{g^{\alpha\beta}\, \partial lg\sqrt{(-g)}/\partial x_\beta\} = -\,\kappa \sum_\sigma T_\sigma{}^\sigma$.

In the second of three papers published by Einstein in November 1915 that led to the final field equations for *general relativity* [Einstein, A. (November 18, 1915). *Erklärung der Perihelbewegung des Merkur aus der allgemeinen Relativitätstheorie.* (Explanation of the Perihelion Motion of Mercury from the General Theory of Relativity.)], Einstein began by noting that from his last two communications, the *gravitational field* in a vacuum *in the absence of matter* had to satisfy, upon properly choosing a reference frame, the *geodesic equations* $\sum_\alpha \partial\Gamma^\alpha_{\mu\nu}/\partial x^\alpha + \sum_{\alpha\beta} \Gamma^\alpha_{\mu\beta}\, \Gamma^\beta_{\nu\alpha} = 0$, where the $\Gamma^\tau_{\mu\nu}$ were defined by the equations $\Gamma^\alpha_{\mu\nu} = -\{\alpha^{\mu\nu}\} = -\sum_\beta g^{\alpha\beta}\, [\beta\, ^{\mu\nu}] = -\tfrac{1}{2}\sum_\beta g^{\alpha\beta}\, (\delta g_{\mu\beta}/\delta x_\nu + \delta g_{\nu\beta}/\delta x_\mu - \delta g_{\mu\nu}/\delta x_\alpha)$. He then considered the case in which a *point mass*, the Sun, was located at the origin of the coordinate system, and noted that the gravitational field this *point mass* produced could be calculated from these equations by means of successive approximations.

He considered cases *when the velocity of the particle was very small compared with the speed of light* and the $g_{\mu\nu}$ differed from the values in an inertial frame under special relativity only by small magnitudes so that small quantities of the second and higher orders could be neglected (his "first aspect of the approximation") and dx_1/ds, dx_2/ds, dx_3/ds could be treated as small quantities, whereas dx_4/ds was equal to 1 (his "second point of view for approximation). Einstein claimed that in the *first approximation* his *assumed solution* $g_{\rho\sigma} = -\delta_{\rho\sigma} - \alpha x_\rho x_\sigma/r^3$ and $g_{44} = 1 - \alpha/r$, satisfied these *geodesic equations*, where $\delta_{\rho\sigma}$ was equal to 1 or 0 if $\rho = \sigma$ or $\rho \# \sigma$, respectively, $r = \sqrt{(x_1^2 + x_2^2 + x_3^2)}$, and α was a constant determined by the mass of the Sun. He noted that according to his *theory of general relativity* $ds^2 = \sum g_{\mu\nu} dx_\mu dx_\nu = 0$, determining the velocity of light, so that light-rays are bent if the $g_{\mu\nu}$ were not constant. From this, Einstein calculated the *deflection of light by the Sun* at a distance Δ, $B = 2\alpha/\Delta = \kappa M/2\pi\Delta$, by substituting $\alpha = \kappa M/4\pi$, from his equation for the *gravitational potential* $\varphi(r) = -\frac{1}{2} \alpha/r = -\kappa/8\pi \int \rho d\tau/r = -\kappa M/8\pi r$. He obtained a value for κ (or α) by substituting the value obtained by equating *his equation for the gravitational potential* $\varphi(r) = -\kappa M/8\pi r$ with the equation for the *gravitational potential under the Newtonian theory* $\varphi(r) = -\frac{1}{2} \alpha/r = -K/c^2 \int \rho d\tau/r = -KM/c^2 r$, so $-\kappa M/8\pi r = -KM/c^2 r$, where K denotes the gravitation constant 6.7×10^{-10}, giving $\kappa = 8\pi K/c^2 = 1.87 \times 10^{-29}$ (using consistent units). This produced a deflection of 1.7 arcseconds, which was the Newtonian result, bringing into line with the Newtonian calculation published by Soldner in 1804.

In order to determine the *orbits of the planets* he calculated the *second approximation* of the last field equation to obtain $\Gamma^\sigma_{44} = -a/2\ x_\sigma/r^3\ (1 - a/r)$. Applying this to the equations of a *mass point in a gravitational field*, $d^2x_\nu/ds^2 = \sum_{\sigma,\tau} \Gamma^\alpha_{\sigma\tau} dx_\sigma/ds\ dx_\tau/ds$, he obtained the equation for the *motion of the planets*, $(dx/d\phi)^2 = 2A/B^2 + a/B^2\ x - x^2 + ax^3$, where ϕ was the angle described by the radius vector between the perihelion and the aphelion, $x = 1/r$, $A = \frac{1}{2}\ (dr^2/ds^2 + r^2\ d\phi^2/ds^2 - a/r)$, and $B = r^2 d\phi/ds$. This *second approximation of the geodesic equation*, from which an equation in the form of Newton's law could be obtained mathematically, differed from the corresponding one in Newtonian theory by an additional term $+ ax^3$, referred to as Einstein's *post-Newtonian expansion*. By integration of the elliptical integral Einstein obtained the *contribution of the additional term*, and deduced that after a complete orbit, the perihelion of Mercury advanced by an additional amount $\varepsilon = 3\pi a/a(1 - e^2)$, or $\varepsilon = 24\pi^3 a^2/T^2 c^2(1 - e^2)$ in terms of the orbital period. In order to obtain a value for ε, it was again necessary to substitute for Einstein's *gravitational constant from Newton's equation for the gravitational potential*. Substitution of values in these equations results in *42.9 arcseconds per Julian century*, appearing to confirm Einstein's claim.

The third of three papers published by Einstein in November 1915 that led to the final field equations for *general relativity* [Einstein, A. (November 25, 1915). *Die Feldgleichungen der Gravitation.* (The Field Equations of Gravitation.)], *was seen to be the defining paper of general relativity*. At long last, Einstein felt that he had found workable *field equations*. Einstein noted that in his previous papers the hypothesis had to be introduced that the *scalar of the energy tensor of matter* disappeared. In this paper he reported that he could do away with this hypothesis *if the energy tensor of matter was inserted into the field equations in a slightly different way.* From the covariant tensor of the second rank $G_{im} = R_{im} + S_{im}$, where $R_{im} = -\sum_l \partial/\partial x_l \{_l^{im}\} + \sum_{lp} \{_\rho^{il}\}\{_l^{mp}\}$ and $S_{im} = \sum_l \partial/\partial x_m \{_l^{il}\} - \sum_{lp} \{_\rho^{im}\}\{_l^{\rho l}\}$, where $\{_i^{im}\} = \frac{1}{2} g^{l\tau} (\partial g_{i\tau}/\partial x_m + \partial g_{m\tau}/\partial x_i - \partial g_{im}/\partial x_\tau)$, the ten generally-covariant equations of the *gravitational field* in spaces *where "matter" was absent* were obtained by setting $G_{im} = 0$. By choosing the frame of reference so that $\sqrt{(-g)} = 1$, S_{im} vanishes because $R_{\mu\nu} = \sum_\alpha \partial\Gamma^\alpha_{\mu\nu}/\partial x^\alpha + \sum_{\alpha\beta} \Gamma^\alpha_{\mu\beta} \Gamma^\beta_{\nu\alpha} = -\kappa T_{\mu\nu}$ and $R_{im} = \sum_l \partial\Gamma^l_{im}/\partial x_l + \sum_{\rho l} \Gamma^l_{i\rho} \Gamma^\rho_{ml} = 0$, where $\Gamma^l_{im} = -\{_l^{im}\}$ are *the "components" of the gravitational field. If "matter" was present* in the space under consideration, its *energy tensor* occurred on the right side of $G_{im} = 0$ or $R_{im} = \sum_l \partial\Gamma^l_{im}/\partial x_l + \sum_{\rho l} \Gamma^l_{i\rho} \Gamma^\rho_{ml} = 0$. Setting $G_{im} = -\kappa (T_{im} - \frac{1}{2} g_{im}T)$, where $T = \sum_{\rho\sigma} g^{\rho\sigma}T_{\rho\sigma} = \sum_{\rho\sigma} T_\rho^\sigma$ was the *scalar of the energy tensor of "matter"*, and specializing the coordinate system, he obtained in place of $G_{im} = -\kappa (T_{im} - \frac{1}{2} g_{im} T)$, $R_{im} = \sum_l \partial\Gamma^l_{im}/\partial x_l + \sum_{\rho l} \Gamma^l_{i\rho} \Gamma^\rho_{ml} = -\kappa (T_{im} - \frac{1}{2} g_{im}T)$, and $(-g)^{1/2} = 1$. Assuming that *the divergence of matter vanished*, the *conservation law of matter and the gravitational field combined* became $\sum_\lambda \partial/\delta x_\lambda (T_\sigma^\lambda + t_\sigma^\lambda) = 0$, where t_σ^λ, the *"energy tensor" of the gravitational field*, was given by $\kappa t_\sigma^\lambda = \frac{1}{2} \delta_\sigma^\lambda \sum_{\mu\nu\alpha\beta} g^{\mu\nu} \Gamma^\alpha_{\mu\beta} \Gamma^\beta_{\nu\alpha} - \sum_{\mu\nu\alpha} g^{\mu\nu} \Gamma^\alpha_{\mu\sigma} \Gamma^\beta_{\nu\alpha}$.

In a letter to Arnold Sommerfeld dated November 28, 1915, immediately after he had submitted his three November 1915 papers, Einstein stated his reasons for abandoning the "Entwurf" field equations and recounted the subsequent developments. He explained that the correct equations were $G_{im} = -\kappa\{T_{im} - \frac{1}{2} g_{im} \sum_{\alpha\beta} (g^{\alpha\beta} T_{\alpha\beta})\}$, and that by choosing the frame of reference in such a way that $\sqrt{-g} = 1$, the equations became $-\sum_l \partial\{^{im}_l\}/\partial x_l + \sum_{\alpha\beta} \{^{i\alpha}_\beta\} \{^{m\beta}_\alpha\} = -\kappa(T^{im} - \frac{1}{2} g_{im}T)$,, where $T = \sum_{\alpha\beta} (g^{\alpha\beta} T_{\alpha\beta})$ was the *scalar of the energy tensor of the "matter"*. He explained that this came from the realization that not $\sum g^{l\alpha} \partial g_{\alpha i}/\partial x_m$ but the related Christoffel symbols $\{^{im}_l\}$ should be regarded as a natural expression for the "*component*" of the *gravitational field*. Einstein also claimed that "not only Newton's theory emerged as the *first approximation*, but also the *perihelion motion of Mercury* (43" per century) as a *second approximation*. For the deflection of light from the sun, the amount was twice as high as before".

In a letter to Einstein dated December 22, 1915, written while he was serving in the war stationed on the Russian front, and subsequently published as a paper [Schwarzschild, K. (1916). *Über das Gravitationsfeld eines Massenpunktes nach der Einstein'schen Theorie.* (On the Gravitational Field of a Point-Mass, according to Einstein's Theory.)], Karl Schwarzschild provided an exact solution to the equation for the *motion of a point* moving along a *geodesic line* where the "components of the gravitational field", Γ, satisfied Einstein's "*field equations*" $\sum_\alpha \partial\Gamma^\alpha_{\mu\nu}/\partial x^\alpha + \sum_{\alpha\beta} \Gamma^\alpha_{\mu\beta} \Gamma^\beta_{\nu\alpha} = 0$, then used this to derive the equation for the *motion of the planets*, $(dx/d\phi)^2 = (1-h)/c^2 + h\alpha/c^2 x - x^2 + \alpha x^3$. Substituting $c^2/h = B^2$ (*not $c^2/h = B$*) and $(1-h)/h = 2A$, this was identical to Einstein's equations $(dx/d\phi)^2 = 2A/B^2 + \alpha/B^2 x - x^2 + \alpha x^3$, for the *motion of the planets*, where ϕ was the angle described by the radius vector between the perihelion and the aphelion, and $x = 1/r$. As

70

with Einstein (November 18, 1915) [*Erklärung der Perihelbewegung des Merkur aus der allgemeinen Relativitätstheorie.*], this did not represent the *weak attractive force of gravitation*, so it was necessary to import Newton's law of gravitation, resulting in what was effectively an extension of the Newtonian result. In March 1916, Schwarzschild was cleared from service due to his sickness and returned to Göttingen. Two months later, on May 11, 1916, Schwarzschild's struggle with pemphigus led to his death at the age of 42.

In March 1916, Einstein published a final consolidation of his various papers - in particular, his three papers in November 1915 [Einstein, A. (1916). *Die Grundlage der allgemeinen Relativitätstheorie.* (The foundation of the general theory of relativity.)]. This was based on the conclusion in his *theory of general relativity* that space and time quantities could not be defined in such a way that spatial coordinate differences could be measured directly with the unit scale, or temporal ones with a normal clock. Einstein assumed that the general laws of nature should be expressed by equations that applied to all coordinate systems not just inertial systems, i.e. were covariant to arbitrary substitutions (generally covariant); and that the *theory of special relativity* was applicable for *infinitely small four-dimensional areas*. He assumed $ds^2 = \sum_{\mu\nu} g_{\mu\nu} \, dx_\mu \, dx_\nu$, where $g_{\mu\nu}$ was the "*fundamental tensor*", which described a curved surface, the *gravitational field*. He introduced the *extension* of the *fundamental tensor* $g_{\mu\nu}$, known as the *Riemann-Christoffel Tensor*, and equated the *equation of motion* of a freely moving body in a frame moving with uniform acceleration relative to the reference frame, i.e. along a *geodetic line* in space time, with the *equation of motion* of a material-point in a *gravitational field*.

Einstein used the *field equations* of forces arising in an accelerated frame in the absence of matter, expressed in terms of the Hamiltonian, to obtain an equation corresponding to the *laws of conservation of momentum and energy*, in terms of the *energy components t_σ^α of the gravitation field*, adding an arbitrary factor -2κ, to obtain
$\kappa t_\sigma^\alpha = \frac{1}{2} \delta_\sigma^\alpha g^{\mu\nu} \Gamma^\lambda_{\mu\beta} \Gamma^\beta_{\nu\lambda} - g^{\mu\nu} \Gamma^\alpha_{\mu\beta} \Gamma^\beta_{\nu\sigma}$, where $\Gamma^\tau_{\mu\nu} = -\frac{1}{2} g^{\tau\alpha} (\partial g_{\mu\alpha}/\partial x_\nu + \partial g_{\nu\alpha}/\partial x_\mu - \partial g_{\mu\nu}/\partial x_\alpha)$. Einstein then introduced *matter* into the *field equations* by adding an *energy-tensor T_σ^α associated with matter*, "corresponding to the density ρ of Poisson's equation $\Delta\varphi = 4\pi\kappa\rho$, where φ is the gravitational potential and ρ is the density of matter", to obtain the *general field equations of gravitation* in the form $\partial/\partial x_\alpha (g^{\sigma\beta}\Gamma^\alpha_{\mu\beta}) = -\kappa\{(t_\mu^\sigma + T_\mu^\sigma) - \frac{1}{2} \delta_\mu^\sigma (t + T)\}$, $(-g)^{1/2} = 1$, or $\partial\Gamma^\alpha_{\mu\nu}/\partial x_\alpha + \Gamma^\alpha_{\mu\beta} \Gamma^\beta_{\nu\alpha} = -\kappa(T_{\mu\nu} - \frac{1}{2} g_{\mu\nu}T)$, with $(-g)^{1/2} = 1$, with the *sum of the energy components of matter and gravitation*, $t_\mu^\sigma + T_\mu^\sigma$ in place of the *energy components* t_μ^σ, where $t = t^\alpha_\alpha$, and $T = T_\mu^\mu$ (Laue's scalar).

He then introduced *Euler's equation of motion for a frictionless adiabatic liquid* in a *relativistic* form in which the *contravariant energy-tensor* of the liquid was
$T^{\alpha\beta} = -g^{\alpha\beta}p + \rho \, dx_\alpha/ds \, dx_\beta/ds$ in an attempt to provide a link between the *stress-energy tensor* defined in his *field equations* and *matter*. However, the force on matter in Euler's

equation is much stronger, has nothing to do with the weak force of gravitational attraction between matter, *and is of opposite sign.*

Einstein then considered cases *where the velocity of the particle was very small compared with the speed of light* and the $g_{\mu\nu}$ differed from the values in an inertial frame under special relativity only by small magnitudes, so that small quantities of the second and higher orders could be neglected (his "first aspect of the approximation") and dx_1/ds, dx_2/ds, dx_3/ds could be treated as small quantities, whereas dx_4/ds was equal to 1 (his "second point of view for approximation). This reduced his *equation of motion of a particle moving along the geodesic line* from $d^2x_\tau/ds^2 = \Gamma^\tau{}_{\mu\nu} \, dx_\mu/ds \, dx_\nu/ds$, where $\Gamma^\tau{}_{\mu\nu} = -\frac{1}{2} \, g^{\tau\alpha} \, (\partial g_{\mu\alpha}/\partial x_\nu + \partial g_{\nu\alpha}/\partial x_\mu - \partial g_{\mu\nu}/\partial x_\alpha)$, to $d^2x_\tau/dt^2 = -\frac{1}{2} \, \partial g_{44}/\partial x_\tau$ ($\tau = 1, 2, 3$), which Einstein considered represented the motion of a material point according to Newton's theory, in which $g_{44}/2$ played the part of the *gravitational potential.*

Under a series of approximations to the *contravariant energy-tensor* of a frictionless adiabatic the liquid, $T^{\alpha\beta}$, all components vanished except $T_{44} = \rho = T$, from which Einstein obtained an equation for the *gravitational potential* in terms of the integral of the density of matter divide by the distance from the center of the matter $\varphi(r) = -\kappa/8\pi \int \rho d\tau/r$, of similar form to Newton's law of gravitation $\varphi(r) = -K/c^2 \int \rho d\tau/r$. *In order to obtain a value for κ, Einstein set these two equations equal* giving $\kappa = 8\pi K/c^2 = 1.87 \times 10^{-29}$ (after correction for units), where K is the gravitation-constant 6.7×10^{-10}, in $\kappa = 8\pi K/c^2 = 1.87 \times 10^{-29}$.

As in Einstein (November 18, 1915) [*Erklärung der Perihelbewegung des Merkur aus der allgemeinen Relativitätstheorie.*], his calculation of the bending of light, was obtained from his approximations for his equation of the *geodetic line* $\sum_\alpha \partial\Gamma^\alpha{}_{\mu\nu}/\partial x^\alpha + \sum_{\alpha\beta} \Gamma^\alpha{}_{\mu\beta} \Gamma^\beta{}_{\nu\alpha} = 0$, where $\Gamma^\alpha{}_{\mu\nu} = -\frac{1}{2} \sum_\beta g^{\alpha\beta} (\delta g_{\mu\beta}/\delta x_\nu + \delta g_{\nu\beta}/\delta x_\mu - \delta g_{\mu\nu}/\delta x_\alpha)$, in which the link to the weak attractive force of gravitation was provided by *Newton's law of gravitation.* He noted that according to his *theory of general relativity* $ds^2 = \sum g_{\mu\nu} dx_\mu dx_\nu = 0$, determining the velocity of light, so that light-rays were bent if the $g_{\mu\nu}$ were not constant. Einstein calculated the *deflection of light by the Sun* at a distance Δ, $B = 2\alpha/\Delta = \kappa M/2\pi\Delta$, by substituting $\alpha = \kappa M/4\pi$, from his equation for the *gravitational potential* $\varphi(r) = -\frac{1}{2} \alpha/r = -\kappa/8\pi \int \rho d\tau/r = -\kappa M/8\pi r$, and setting $\kappa = 8\pi K/c^2 = 1.87 \times 10^{-29}$. Consequently, as before, his computed value for the bending of light was the Newtonian value. He restated his formula for the addition to the precession of the perihelion of Mercury, but did not provide the derivation. Why anyone gave credence to this is a mystery. By 1921 Einstein was already moving his research interests into superseding general relativity.

In November 1916, in response to Lorentz and Hilbert's success in presenting the *theory of general relativity* in a comprehensive form by deriving its equations from a single *variational principle*, Einstein published his own version, making as few assumptions about the constitution of matter as possible [Einstein, A. (November, 1916). *Hamiltonsches*

Prinzip und allgemeine Relativitätstheorie. (Hamilton's principle and general relativity.)] Assuming the *gravitational field* to be described as usual by the tensor of the $g_{\mu\nu}$, and *matter* (inclusive of the electromagnetic field) by an arbitrary number of space-time functions $q_{(\rho)}$, and $\mathscr{H}$ to be a function of the $g^{\mu\nu}$, $g_\sigma^{\mu\nu}$ ($= \partial g^{\mu\nu}/\partial x_\sigma$), $g_{\sigma\tau}^{\mu\nu}$ ($= \partial^2 g^{\mu\nu}/\partial x_\sigma \partial x_\tau$), $q_{(\rho)}$ and $q_{(\rho)\alpha}$ ($= \partial q_{(\rho)}/\partial x_\alpha$), the *variational principle* δ $\{\int \mathscr{H}\, d\tau\} = 0$ provided as many differential equations as there were functions $g_{\mu\nu}$ and $q_{(\rho)}$. Einstein then assumed that $\mathscr{H} = \mathbf{G} + \mathbf{M}$, where $\mathbf{G}$ depended only upon $g^{\mu\nu}$, $g_\sigma^{\mu\nu}$, $g_{\sigma\tau}^{\mu\nu}$, and $\mathbf{M}$ only upon $g^{\mu\nu}$, $q_{(\rho)}$, $q_{(\rho)\alpha}$, obtaining the *field equations of gravitation and matter* in the form
$\partial/\partial x_\alpha \{\partial \mathbf{G} \cdot /\partial g_\alpha^{\mu\nu}\} - \partial \mathbf{G} \cdot /\partial g^{\mu\nu} = \partial \mathbf{M}/\partial g^{\mu\nu}$ and $\partial/\partial x_\alpha \{\partial \mathbf{M}/\partial q_{(\rho)\alpha}\} - \partial \mathbf{M}/\partial q_{(\rho)} = 0$. Einstein next assumed that $ds^2 = g_{\mu\nu}\, dx_\mu dx_\nu$ was an invariant, which fixed the transformational character of the $g_{\mu\nu}$ from which he derived his field equations in the form
$\partial/\partial x_\alpha (\partial \mathbf{G}/\partial g_\alpha^{\mu\nu}\, g^{\mu\nu}) = - (\mathbf{T}_\sigma^\nu + \mathbf{t}_\sigma^\nu)$, where $\mathbf{T}_\sigma^\nu = - \partial \mathbf{M}/\partial g^{\mu\nu}$ and
$\mathbf{t}_\sigma^\nu = - (\partial \mathbf{G} \cdot /\partial g_\alpha^{\mu\nu}\, g_\alpha^{\mu\nu} + \partial \mathbf{G} \cdot /\partial g^{\mu\nu}\, g^{\mu\nu})$. From the *field equations* of gravitation *alone*, using the postulate of general covariance, Einstein obtained $\partial/\partial x_\nu (\mathbf{T}_\sigma^\nu + \mathbf{t}_\sigma^\nu) = 0$, which he claimed expressed *the conservation of the momentum and the energy*, where $\mathbf{T}_\sigma^\nu$ were *the components of the energy of matter*, and $\mathbf{t}_\sigma^\nu$ *the components of the energy of the gravitational field*.

In May 1918 [Einstein, A. (1918). *Prinzipielles zur allgemeinen Relativitätstheorie.* (Principles of the general theory of relativity.)], Einstein proposed a new foundation for general relativity, replacing parts of the foundation laid in his paper of March, 1916. In December 1920 [Einstein, A. (December, 1920). *Antwort auf vorstehende Betrachtung.* (Answer to the above considerations.)], Einstein responded to a question raised by Ernest Reichenbächer (To What Extent Can Modern Gravitational Theory Be Established without Relativity?). Einstein recognized that a theory of gravitation could also be established and justified without the principle of relativity but offered arguments in favor of a relativistic theory.

In March 1921, Einstein commented on Hermann Weyl's attempt to supplement the *general theory of relativity* by adding a further condition of invariance [Einstein, A. (1921). *Eine naheliegende Ergänzung des Fundaments der allgemeinen Relativitätstheorie.* (On a natural addition to the foundation of the general theory of relativity.)]. Weyl's theory was based on two ideas: (1) the *ratios* of components $g_{\mu\nu}$ of the *gravitational potential* have a far more fundamental physical meaning than the components themselves, to which Einstein raised the question "Can the theory of relativity be modified by the assumption that not the quantity ds itself, but only the equation $ds^2 = 0$ has an invariant meaning? (2) Weyl's second idea was related to the method of generalization of the Riemannian metric and to the physical interpretation of the newly arising quantities φ_ν in it. Riemannian geometry contains two assumptions: I. *The existence of transferable measuring rods*. II. *The independence of their length from the path of transfer*. Weyl's generalization of Riemann's

73

metric retained (I) but dropped (II). He allowed the measured length of a measuring rod to depend upon its path of transfer by means of an integral extended over the path of transfer; in general, the integral $\int \varphi_v \, dx_v$ depended on this path where the φ_v were *space functions* which, consequently, codetermined the metric. In the physical interpretation of the theory, these were then identified with the *electromagnetic potentials*. Einstein raised a second question "Under these circumstances, one can ask if a distinct theory can be obtained by dropping from the beginning not only Weyl's assumption (II), but also assumption (I) about the existence of transferable measuring rods (and clocks, resp.)".

In his effort to formulate such a theory, Einstein asked his colleague Wirtinger in Vienna if there was a *generalization of the equation of a geodesic line* such that only the ratios of the $g_{\mu v}$ played a role. Wirtinger showed how such a theory could be obtained starting out from only the invariant meaning of the equation $ds^2 = g_{\mu v} \, dx_\mu \, dx_v = 0$ without using the concept of distance ds, i.e. *without using measuring rods or measuring clocks*. Weyl had previously shown that the tensor $H_{iklm} = R_{iklm} - 1/(d-2) \, g_{il}R_{km} + g_{km}R_{il} - g_{im}R_{kl} - g_{kl}R_{im} + 1/(d-1)(d-2)(g_{il}g_{km} - g_{im}g_{kl})R$ was a Weyl tensor of weight 1, where R_{iklm} was the *Riemann curvature tensor*, and R_{km} the tensor of rank 2 that resulted from the previous one by means of one contraction; R was the scalar resulting from one further contraction, and d was the number of dimensions; a Weyl tensor (of weight n) is a Riemann tensor in which the value of a tensor component is multiplied by λ^n if $g_{\mu v}$ is replaced by $\lambda g_{\mu v}$, where λ is an arbitrary function of the coordinates. The desired generalization of the *geodesic line* was then given by the equation $\delta \int d\sigma\} = 0$, where $d\sigma^2 = Jg_{\mu v} \, dx_\mu \, dx_v$ if J was a Weyl invariant of weight −1.

Einstein's 1921 Princeton lectures, [Einstein, A. (1922). *Vier Vorlesungen über Relativitätstheorie: gehalten im Mai 1921 an der Universität Princeton* (The Meaning of Relativity: Four Lectures Delivered at Princeton University, May 1921.)] have been assumed to have superseded the 1916 review article [*Die Grundlage der allgemeinen Relativitätstheorie.* (The foundation of the general theory of relativity.)] as Einstein's authoritative exposition of his theory. Lecture 1 described space and time in pre-relativity physics. Lecture 2 addressed Einstein's theory of special relativity, and lectures 3 and 4 presented Einstein's *theory of general relativity*. In lecture 3, Einstein presented his *theory of gravity* based on the equivalence of gravity and a uniformly accelerated reference frame. In addressing the consequences for his *theory of special relativity* of introducing an accelerated reference frame, Einstein tried to come to terms with, or explain away, the Ehrenfest paradox, whilst preserving the metric of *special relativity* by treating an infinitesimal area of the curved surface as flat. This led him to express the invariant ds between neighboring points linearly in terms of the co-ordinate differentials dx_v in the form $ds^2 = g_{\mu v} \, dx_\mu \, dx_v$, where the functions $g_{\mu v}$ described, with respect to the arbitrarily chosen system of co-ordinates, the metrical relations of the *space-time continuum* and also the *gravitational field*. From this he determined that his *theory of general relativity* required a generalization of the theory of invariants and the theory of tensors. In lecture 4, Einstein applied this mathematical apparatus to the formulation of his *theory of general relativity*.

In addressing the generalization of the motion of a material point on which no force acted Einstein noted that the simplest generalization of a straight line was the *geodesic*, and assumed that, in accordance with the *principle of equivalence*, the motion of a material particle *under the action of only inertia and gravity* was described by the equation $d^2x_\mu/ds^2 + \Gamma^\mu{}_{\alpha\beta} \, dx_\alpha/ds \, dx_\beta/ds = 0$, in which, *by analogy with Newton's equations, the first term represented inertia and the second the gravitational force*. In a first approximation the equation of motion became $d^2x_\mu/ds^2 = 0$, and in a second approximation, he put $g_{\mu\nu} = -\delta_{\mu\nu} + \gamma_{\mu\nu}$, where the $\gamma_{\mu\nu}$ were small of the first order. Both terms of his *equation of motion* were then small of the first order; and neglecting terms that, relative to these, were small of the first order, he obtained $\Gamma^\mu{}_{\alpha\beta} = -\delta_{\mu\nu} \left[\begin{smallmatrix}\alpha\beta\\\sigma\end{smallmatrix}\right] = -\left[\begin{smallmatrix}\alpha\beta\\\mu\end{smallmatrix}\right] = \frac{1}{2} \left(\partial\gamma_{\alpha\beta}/\partial x_\mu - \partial\gamma_{\alpha\mu}/\partial x_\beta - \partial\gamma_{\beta\mu}/\partial x_\alpha\right)$. Then in the case where *the velocity of the mass point was very small compared to the speed of light*, and the gravitational field was assumed to depend on time so weakly that the derivatives of the $\gamma_{\mu\nu}$ by x_4 could be neglected, the *equation of motion* (for $\mu = 1, 2, 3$) reduced to $d^2x_\mu/dl^2 = \partial/\partial x_\mu \, (\gamma_{44}/2)$, which Einstein claimed was identical to *Newton's equation of motion* of a point mass in the gravitational field *if one identified ($\gamma_{44}/2$) with the potential of the gravitational field*.

As in his March, 1916 paper [*Die Grundlage der allgemeinen Relativitätstheorie*. (The foundation of the general theory of relativity.)], Einstein then drew on Poisson's equation on the grounds that this was based on the idea that the gravitational field arose from the density ρ of ponderable matter, substituting the *tensor of the energy density* for the scalar of the *mass density*. He introduced a covariant tensor $T_{\mu\nu}$ of the second rank, the "*energy tensor of matter*", which combined the energy density of the electromagnetic field and that of ponderable matter. The *momentum and energy theorem* was then expressed by the fact that the divergence of this tensor disappeared, $\partial T_{\mu\nu}/\partial x_\nu = 0$, so that in the *theory of general relativity* $0 = \partial\mathfrak{C}_\sigma{}^\alpha/\partial x_\alpha - \Gamma^\alpha{}_{\sigma\beta} \, \mathfrak{C}_\alpha{}^\beta$, where ($T_{\mu\nu}$) denoted *the covariant energy tensor of matter*, and $\mathfrak{C}_\sigma{}^\nu$ the corresponding *mixed tensor density*. By analogy with Poisson's equation, he looked for a differential tensor based on Riemann's tensor, following Wehl's suggestion, and from this he obtained the *law of the gravitation field*, $R_{\mu\nu} - \frac{1}{2} g_{\mu\nu}R = -\kappa \, T_{\mu\nu}$, where R_{iklm} was the *Riemann curvature tensor*, R_{km} the tensor of rank 2 that resulted from the previous one by means of one contraction; R was the scalar resulting from one further contraction, and κ denoted *a constant that was related to the gravitational constant of Newton's theory*. Transforming this by multiplying by $g_{\mu\nu}$, and summing over μ and ν, Einstein obtained $R = \kappa g_{\mu\nu}T_{\mu\nu} = \kappa T$. Applying another approximation and setting $T^{\mu\nu} = \sigma\, dx_\mu/ds\; dx_\nu/ds$ and $ds^2 = g_{\mu\nu}\, ndx_\mu\, dx_\nu$, where σ was the density at rest, i.e. the density of ponderable matter, Einstein obtained $\gamma_{11} = \gamma_{22} = \gamma_{33}$ $= -\kappa/4\pi \int \sigma\, dV_0/r$, $\gamma_{44} = + \kappa/4\pi \int \sigma\, dV_0/r$, while all the other $\gamma_{\mu\nu}$ vanished. The last of these equations gave $d^2x/dt^2 = \kappa c^2/8\pi\; \partial/\partial x_\mu \int \sigma\, dV_0/r$, which was in a similar form to Newton's *equation of gravitation.*

In order to apply this to calculations, as previously, *Einstein introduced a link to the weak gravitational attraction between matter by setting his equation to be equal to Newton's equation of gravity*, and obtained a value for $\kappa = 8\pi\, K/c^2 = 1.86.\ 10^{-27}$, where he put $K = 6.67.\ 10^{-8}$. Using consistent units, $K = 6.67.\ 10^{-10}$, and this becomes $\kappa = 1.8736 \times 10^{-29}$ km kg^{-1}. Einstein assumed that in his theory of general relativity the velocity of light was also everywhere the same relative to an inertial system, so $ds^2 = 0$ and $\sqrt{(dx_1{}^2 + dx_2{}^2 + dx_3{}^2)}/dl$ $= 1 - \kappa/4\pi \int \sigma\, dV_0/r$, from which he concluded that a ray of light passing at a distance Δ from the Sun was deflected by $\alpha = \kappa M/4\pi$, equal to 1.7 arcseconds, which was the *Newtonian calculation.* Einstein then addressed the *motion of the perihelion of the planet Mercury.* Instead of deriving this by successive approximations from his field equations, as in his previous papers, he used the *principle of variation* to obtain $\varepsilon = 24\pi^3 a^2/T^2 c^2(1 - e^2)$. This was the first time that Einstein explained that his equations were in *radians per revolution.*

In his South American Travel Diary for his visit to Argentina, Uruguay and Brazil (March 5 – May 11, 1925), Einstein noted on March 17, 1925, the day that he crossed the Equator, "I have become convinced that $R_{ik} - \tfrac{1}{4}\, g_{ik}\, R = T_{ik(el)}$ is not the right thing. Conviction about the impossibility of field theory in its present meaning is strengthening."

Almost the entire development of Einstein's theory of general relativity, which took place between 1907 and 1922, before, during and after the First World War, was conducted in German by Einstein, and other German and German speaking theoretical physicists and mathematicians. Although English translations of Einstein's contributions are available in print form and on the internet in *The Collected Papers of Albert Einstein*, they are scattered across several volumes, and translations of many of the other contributions are not. The author provided many of the translations with the help of Microsoft translator.

Following the publication of Einstein's *theory of general relativity*, alternative *theories of gravity*, including Einstein's theory, were classified in terms of various parameters, including the γ and β, resulting in the early 1970s in the *Parametrized Post-Newtonian (PPN) formalism.* Between March 2011 and August 2014, an attempt was made to obtain evidence in support of Einstein's *theory of general relativity* based on the precession of Mercury's perihelion, using more accurate data obtained from the MESSENGER spacecraft during its orbital phase. However, despite the expenditure of millions of dollars on specialized satellites orbiting Mercury and extremely accurate ranging equipment, *evidence in support of Einstein's theory based on the precession of Mercury's perihelion proved elusive.*

The current explanation of the *anomalous precession of the perihelion of Mercury*, not predicted by a purely Newtonian gravity field, assumes that it is primarily due (42.97 arcseconds per century) to the *gravitoelectric effect*, or velocity-dependent acceleration, that a moving object near a massive, non-axisymmetric, rotating object such as the Sun will experience due in part to the Sun's oblateness. The remaining part, the *Lense–Thirring (gravitomagnetic) precession*, is a much smaller effect (– 0.002 arcseconds per century), induced by the *gravitomagnetic field* of the Sun on planetary orbital motion, in which, *in the weak-field and slow-motion approximation*, the Einstein field equations of general relativity get linearized resembling the Maxwellian equations of electromagnetism.

These effects are assumed to be relativistic, but there are also *non-relativistic* effects: the aliasing classical precessions induced by a host of classical orbital perturbations of gravitational origin by the Sun's oblateness, the multipolar expansion of the Sun's gravitational potential, and the classical secular N-body precessions, which are of the same order of magnitude or much larger than the Lense-Thirring precession.

The Jet Propulsion Laboratory estimates [Park, R. S., Folkner, W. M., Konopliv, A. S., Williams, J. G., Smith, D. E., & Zuber, M. T. (2017). *Precession of Mercury's Perihelion from Ranging to the MESSENGER Spacecraft.*] of the so-called *gravitoelectric* and *gravitomagnetic (Lense–Thirring)* effects published in March 2017 were not based on calculating these effects but from statistical analysis of satellite observations based on the *parametrized post-Newtonian formulation* of Einstein's equation. The *perihelion precession due to the gravitomagnetic effect* (LT) was computed by comparing the nominal ephemeris with an integration performed with the solar angular momentum set to zero; the precession rate due to *solar oblateness* was computed in the same manner; and the *precession due to the gravitoelectric effect* (GE) was essentially a residual, computed by comparing the precession from the integration *with the speed of light essentially infinite*, then subtracting the effect due to LT and planetary *gravitomagnetic* (GM) contributions in the PPN formulation. Table 3, ("The breakdown of estimated contributions to the precession of perihelion of Mercury"), shows how little progress there has been since Einstein's lectures in 1922 and Clemence's tabulation in 1947, or, indeed, since 1704, the date when Newton's *Opticks* was published. *Einstein's equation for the anomalous precession of the perihelion of Mercury does not represent the gravitoelectric effect* that a moving object near a massive, non-axisymmetric, rotating object such as the Sun will experience due to the Sun's oblateness, nor the effects of N-body interactions, or the aliasing effects induced by a host of classical orbital perturbations of gravitational origin.

The analysis in this volume concludes that whilst Newton enumerated laws of nature based on observation and experiments, including his universal law of gravitation, Einstein pursued theories, based on principles and assumptions, in search of evidence. A detailed examination of Einstein's *theory of general relativity* reveals that it is not a *theory of gravity*; it is a *relativistic* theory about the *effects* of gravitation, or more strictly, of a uniformly accelerated reference frame. There is nothing in any version of his theory that represents or explains or provides any connection to the weak attractive gravitational force between matter. In order to make calculations with his theory, Einstein had to import Newton's empirical law of gravitation. Consequently, the only evidence that Einstein could provide for his theory of general relativity was effectively Newtonian.

We are no further forward in understanding the origin of this fundamental force on which the existence of life and the universe depends. Whilst Einstein's and others' objectives in removing a preferred reference frame and the existence of an ether from physics were admirable intentions, Einstein's subsequent fixation on the constancy of the speed of light, or some form of invariant space-time, in the face of reasonable alternatives, such as Ritz's emission theory on which quantum electrodynamics is founded, was not.

In the light of the continued failure of Einstein's efforts to overcome the main objections to his *theory of special relativity* - the Ehrenfest paradox, and its failure to explain the observed Doppler redshift and blueshift of light – or to provide any evidence for it, and in the absence of any supportive evidence for his *theory of general relativity*, both theories must be rejected.

What is needed is needed is a *non-relativistic* classical analysis based on a more accurate model of the Sun than the point mass model employed by Newton and Einstein, as Newton would have probably provided in 1687 had he been aware of the structure and composition of the Sun. A fundamental theoretical explanation for the weak attractive gravitational force between matter is also long overdue.

5   Gravity March 11, 2024

What is gravity?

But is this such a simple law? What about the machinery of it? All we have done is to describe how the Earth moves around the Sun, but we have not said what makes it go. Newton made no hypothesis about this; was satisfied to find what it did without getting into the machinery of it. No one has since given any machinery. ... Why can we use mathematics to describe nature without a mechanism behind it? No one knows. We have to keep going because we find out more that way.
—Richard Feynman. (1963). *The Feynman Lectures on Physics*, Volume I, p. 7-9.

This book picks up where my last book left off. Underwood, T. G. (2023). *General Relativity*: "Conclusion. ... A detailed examination of Einstein's *theory of general relativity* reveals that it is not a *theory of gravity*; it is a *relativistic* theory about the *effects* of gravitation, or more strictly, of a uniformly accelerated reference frame.

[*Gravity*, also known as *gravitation* or a *gravitational interaction*, is a mutual attraction between all particles with mass.]

There is nothing in any version of this theory that represents or explains or provides any connection to the weak attractive gravitational force between matter. We are no further forward in understanding the origin of this fundamental force. ... Einstein's theory of general relativity attempted to extend his theory of special relativity beyond space and time, to include matter and gravitational fields. Gravitation was introduced through the "equivalence principle", the equivalence of the *outcome* of the force of gravity and the acceleration of matter, first recognized in Newton's *Principia*. ... In order to make calculations with his theory, Einstein had to import Newton's law of gravitation, which itself is an empirical law with no fundamental foundation. ... A fundamental theoretical explanation for the weak attractive gravitational force between matter is also long overdue."

This volume attempts to address this omission. Without gravity there would be no Milky Way galaxy, no Solar System, no Sun, no planet Earth, no life. These all depend on the weak attractive force of gravity magnified in large dense masses. Yet we know less about gravity than any other force. The purpose of this book is to try to fill that void.

Part I reviews contributions to the understanding of the effects of gravity, starting with an extract from Johannes Kepler's seminal book [Kepler, J. (1609). *Astronomia Nova ΑΙΤΙΟΛΟΓΗΤΟΣ seu physica coelestis, tradita commentariis de motibus stellae Martis ex observationibus G.V. Tychonis Brahe.* (New Astronomy, reasoned from Causes, or Celestial Physics, Treated by Means of Commentaries on the Motions of the Star Mars, from the Observations of the noble Tycho Brahe)], published in (1609). This contains the results of the astronomer Johannes Kepler's ten-year-long investigation of the motion of Mars. One of the most significant books in the history of astronomy, the *Astronomia Nova* provided strong arguments for heliocentrism and contributed valuable insight into the movement of the planets. This included *the first mention of the planets' elliptical paths* and the change of their movement to the movement of free-floating bodies as opposed to objects on rotating spheres. It records the discovery of *the first two of the three principles known today as Kepler's laws of planetary motion; (1) That the planets move in elliptical orbits with the Sun at one focus*, and (2) *That the speed of the planet changes at each moment such that the time between two positions is always proportional to the area swept out on the orbit between these positions*.

This is followed by an extract from Kepler, J. (1619). *Harmonices mundi libri V*. (The Five Books of The Harmony of the World.), published in 1619, in which Kepler discusses harmony and congruence in geometrical forms and physical phenomena. The final section of this work relates his discovery of his "*third law of planetary motion*", a decade after the publication of the *Astronomia nova*. He found that (3) *The ratio of the cube of the length of the semi-major axis of each planet's orbit, to the square of time of its orbital period, is the same for all planets*.

Then it is Sir Isaac Newton's turn. In July, 1687, Newton published *Philosophiæ Naturalis Principia Mathematica*. (The Mathematical Principles of Natural Philosophy), a work in three books written in Latin. The *Principia* includes Newton's three laws of motion, laying the foundation for classical mechanics; Newton's law of universal gravitation; and a derivation of Johannes Kepler's laws of planetary motion (which Kepler had first obtained empirically).

In *Book I: The Motion of Bodies*. [(*De motu corporum*.)], Newton addresses the motion of bodies attracted to each other by centripetal forces. He proposes that if two bodies, attracting each other with forces reciprocally proportional to the squares of their distance, revolve about their common center of gravity; that the principal axis of the ellipse which either of the bodies describes by this motion about the other, will be to the principal axis of the ellipse, which the same body may describe in the same periodical time about the other body. Of the motions of bodies tending to each other with centripetal forces, he proposes that two bodies attracting each other mutually, describe similar figures about their common center of gravity, and about each other mutually. Part of the contents originally planned for the first book was divided out into a second book, *Book II. Part 2, The Motion of Bodies*. [(*De motu corporum*.)] which largely concerns motion through resisting mediums.

In *Book III: Of the System of the World*, [(*De mundi systemate* .)], Newton notes that the centripetal force which arises between planets is the same as the gravitational force attracting matter to the Earth and focusses on gravitational attraction. He observes that the circumjovial planets, by radii drawn to Jupiter's center, describe areas proportional to the times of description; and that their periodic times, the fixed stars being at rest, are in the sesquiplicate proportion of their distances from, its center, etc. That the fixed Stars being at rest, the periodic times of the five primary Planets, and (whether of the Sun about the Earth, or) of the Earth about the Sun, are in the sesquiplicate proportion of their mean distances from the Sun. He noted that this proportion, first observed by Kepler, is now accepted by all astronomers.

> [The *sesquiplicate* ratio of given terms is the ratio between the square roots of the cubes of those terms.]

He further observes that then the primary Planets, by the areas, which they describe by radii drawn to the Sun, are proportional to the times of description; and that the Moon by a radius drawn to the Earth's center, describes an area proportional to the time of description. From this he proposes that the forces by which the circumjovial planets are continually drawn off from rectilinear motions, and retained in their proper orbits, tend to Jupiter's center; and are reciprocally as the squares of the distances of the places of those planets from that center; that the forces by which the primary planets are continually drawn off from rectilinear motions, and retained in their proper orbits, tend to the Sun; and are reciprocally as the squares of the distances of the places of those planets from the Sun's center; and that the force by which the Moon is retained in its orbit tends to the Earth; and is reciprocally as the square of the distance of its place from the Earth's center. He then proposes that that the Moon gravitates towards the Earth, and by the force of gravity is continually drawn off from a rectilinear motion, and retained in its orbit; that the circumjovial planets gravitate towards Jupiter; the circumsaturnal towards Saturn; the circumsolar towards the Sun; and by the forces of their gravity are drawn off from rectilinear motions, and retained in curvilinear orbits; and concludes that the force which retains the celestial bodies in their orbits has been hitherto called centripetal force; but it being now made plain that it can be no other than a gravitating force, we shall hereafter call it gravity. He then proposes that all bodies gravitate towards; every Planet and that the Weights of bodies towards any the same Planet, at equal distances from the center of the Planet, are proportional to the quantities of matter which they severally contain; and that there is a power of gravity tending to all bodies, proportional to the several quantities of matter which they contain; and that the force of gravity towards the several equal particles of any body, is reciprocally as the square of the distance of places from the particles; as appears from Cor. 3, Prop. LXXIV, Book I.

81

In Proposition VI, Newton provides his definition of *gravitational mass*, and in Proposition VII, together with its corollary 2, Newton restates his *universal law of gravitation*.

Proposition VI. Theorem VI. *That all bodies gravitate towards; every Planet and that the Weights of bodies towards any the same Planet, at equal distances from the center of the Planet, are proportional to the quantities of matter which they severally contain.*

Proposition VII. Theorem VII. *That there is a power of gravity tending to all bodies, proportional to the several quantities of matter which they contain.*
…
Corollary 2. The force of gravity towards the several equal particles of any body, is reciprocally as the square of the distance of places from the particles; as appears from Cor. 3, Prop. LXXIV, Book I.

However, while Newton was able to articulate his *Law of Universal Gravitation* and verify it experimentally, he could only calculate the relative gravitational force in comparison to another force. It was not until Henry Cavendish's verification in 1798 [Cavendish, H. (1798). *Experiments to Determine the Density of Earth.*] of the Gravitational Constant, that the Law of Universal Gravitation received its final form,

$$F = GMm/r^2 = 6.674 \times 10^{-11}\ Mm/r^2\ N\ (\textit{SI units}),$$

where F represents the force in Newtons, M and m represent the two masses in kilograms, and r represents the separation in meters. G represents the gravitational constant, which has a value of $6.674 \times 10^{-11}\ N\ (m/kg)^2$.

Because of the magnitude of G, gravitational force is very small unless large masses or short distances are involved. Cavendish used apparatus, consisting of a wooden arm, 6 feet long, made so as to unite great strength with little weight. This arm was suspended in a horizontal position, by a slender wire 4.0 inches long, and to each extremity is hung a leaden ball, about 2 inches in diameter; and the whole is enclosed in a narrow wooden case, to defend it from the wind. As no more force is required to make this arm turn round on its center, than what is necessary to twist the suspending wire, it is plain, that if the wire is sufficiently slender, the most minute force, such as the attraction of a leaden weight a few inches in diameter, will be sufficient to draw the arm sensibly aside.

Part I also describes *Gauss's law for gravity*, named after Carl Friedrich Gauss, also known as Gauss's flux theorem for gravity, which is a law of physics that is equivalent to Newton's law of universal gravitation; and *Poisson's equation*, which provides the *potential field* caused by a given *mass density distribution* from which the *gravitational (force) field* can be calculated.

Gauss's law for gravity states that *the flux (surface integral) of the gravitational field over any closed surface is proportional to the enclosed mass.* It is often more convenient to work from than Newton's law.

The form of *Gauss's law for gravity* is mathematically similar to Gauss's law for electrostatics, one of Maxwell's equations. *Gauss's law for gravity* has the same mathematical relation to Newton's law that Gauss's law for electrostatics bears to Coulomb's law. This is because both Newton's law and Coulomb's law describe inverse-square interaction in a 3-dimensional space.

The *gravitational field* g (also called gravitational acceleration) is a vector field – a vector at each point of space (and time). It is defined so that the *gravitational force* experienced by a particle is equal to the mass of the particle multiplied by the *gravitational field* at that point.

The left-hand side of integral form of Gauss's law for gravity is called the *flux of the gravitational field*. According to the law it is always negative (or zero), and never positive. This can be contrasted with Gauss's law for electricity, where the flux can be either positive or negative. The difference is because charge can be either positive or negative, while mass can only be positive. *Gravitational flux* is a surface integral of the *gravitational field* over a closed surface, analogous to how magnetic flux is a surface integral of the magnetic field.

Poisson's equation is an elliptic partial differential equation of broad utility in theoretical physics, first published in 1813. The solution to Poisson's equation is the *potential field* caused by a given electric charge or mass density distribution; with the potential field known, one can then calculate the electrostatic or gravitational (force) field. If the mass density is zero, Poisson's equation reduces to Laplace's equation. The corresponding Green's function can be used to calculate the potential at distance r from a central point mass m (i.e., the fundamental solution). In three dimensions the *potential* is

$$\Phi(r) = -\,Gm/r$$

which is equivalent to Newton's law of universal gravitation.

Part I also includes the annotated copy of Einstein's March, 1916, paper [Einstein, A. (1916). *Die Grundlage der allgemeinen Relativitätstheorie*. (The foundation of the general theory of relativity.)], Einstein's final consolidation of his various papers on the subject, from Underwood, T. G. *General Relativity*, pp. 347-409. This shows how Einstein obtained an equation for the *gravitational potential* in terms of the integral of the density of matter divide by the distance from the center of the matter and substituted Newton's law of gravitation; and how his calculation of the bending of light, was obtained from his approximations for his equation of the *geodetic line*, in which the link to the weak attractive force of gravitation was also provided by *Newton's law of gravitation*. Consequently, his computed value for the bending of light was the Newtonian value.

A detailed examination of Einstein's *theory of general relativity* reveals that it is not a *theory of gravity*; it is a *relativistic* theory about the *effects* of gravitation, or more strictly, of a uniformly accelerated reference frame. There is nothing in any version of this theory that represents or explains or provides any connection to the weak attractive gravitational force between matter. In order to make calculations with his theory, Einstein had to import Newton's empirical law of gravitation. We are no further forward in understanding the origin of this fundamental force. Whilst Einstein's and others' objectives in removing a preferred reference frame and the existence of an ether from physics were admirable intentions, Einstein's subsequent fixation on the constancy of the speed of light, or some form of invariant space-time, in the face of reasonable alternatives, such as Ritz's emission theory on which quantum electrodynamics is founded, was not.

Finally, Part I includes an extract from Einstein, A. (1917). *Kosmologische Betrachtungen zur allgemeinen Relativitätstheorie*. (Cosmological Considerations in the General Theory of Relativity.), which describes Einstein's struggles with supplementing the *relativistic differential equations* by *limiting conditions* at *spatial infinity* in order to regard the universe as being of infinite spatial extent. In his treatment of the planetary problem, he chose these limiting conditions on the basis of the assumption that it is possible to select a system of reference so that at spatial infinity all the gravitational potentials $g_{\mu\nu}$ become constant, but it was by no means evident that the same limiting conditions could be applied to larger portions of the physical universe. Einstein attempts to resolve this using a method analogous to the extension of Poisson's equation used in the *non-relativistic case*, by adding to the left-hand side of field equation, the fundamental tensor $g_{\mu\nu}$, multiplied by a universal constant, $-\lambda$. As he noted, "*we admittedly had to introduce an extension of the field equations of gravitation which is not justified by our actual knowledge of gravitation*".

Part II addresses "What is Gravity?". It begins by reviewing Einstein's unsuccessful attempts at producing a *classical unified field theory* between 1923 until he died in 1955, during which time Einstein published 31 papers on a unified theory of electromagnetism and gravity. An extract of the first of these, which he published jointly with Jacob Grommer is included [Einstein, A. & Grommer, J. (1923). *Proof of the Non-Existence of an Everywhere Regular, Centrally Symmetric Field According to the Field Theory of Kaluza.*].

Part II also includes Heinrich Weyl's attempt in 1929 to incorporate Dirac theory into the scheme of *general relativity* by introducing *gauge invariance* of *theory of coupled electromagnetic potentials* and Dirac *matter waves* [Weyl, H. (1929). *Elektron und Gravitation.* (Electron and gravity.)] Weyl claimed that the barrier which hems progress of quantum theory is quantization of the field equations.

It also notes Einstein, A., Podolsky, B. & Rosen, N. May, 1935 paper, [*Can Quantum-Mechanical Description of Physical Reality Be Considered Complete?*] which suggests that the description of reality as given by a wave function in quantum mechanics is not complete.

But then it moves on to consider *quantum entanglement,* or some form of entanglement between matter, as a potential source of gravity and to examine the origin of gravity according to the Big Bang theory. In an attempt to answer the question: "Why does matter attract matter?", facts related to gravity and other forces of nature were are listed and reviewed. These include the following:

(1)   The gravitational attractive force between two bodies obeys an inverse square law as if it were a radiated force in three-dimensional space;

(2)   The gravitational attractive force between two bodies is proportional to the product of the masses, m_1 and m_2, of the bodies being attracted;

(3)   The gravitational attractive force is additive in the sense that each of the constituents (one atom or one molecule) of one body are attracted by the each of the constituents of the other body in proportion to the product of each of their masses;

(4)   The force of attraction between two objects with opposite charges also obeys an inverse square law;

(5)   The gravitational force is about 10^{36} times weaker than the electric force. As both forces are subject to inverse square laws, this relationship will apply at all distances;

(6)   The net inward gravitational force on 1 kg of matter at a distance s from the center of a planet or a star with radius r and density ρ is given by

$$F_{net} = 6.674 \times 10^{-11} \, 4/3 \, \pi(2s - r) \, \rho \text{ N};$$

(7)   There appears to be no equivalent gravitational repulsive force;

(8)   The gravitational field of each body is present at every other body;

(9)   According to Big Bang theory, the force of gravity separated from the other forces as the universe's temperature fell, immediately prior to the cosmic inflation during which the universe grew exponentially by a factor of at least 10^{78};

(10)   According to Newton's universal law, the inward gravitational force F on a mass m at the at the surface of the universe,

$$F = GmM/r^2,$$

and the change in the inward gravitational force F on a mass m at the at the surface of the universe with the increase of the radius, due to the expansion of the universe, is inversely proportional to the cube of the radius of the universe.

$$dF/dr = -2 \, GmM/r^3;$$

(11)   One way of looking at the expansion of the universe is as the expansion of the metric which determines the size of the universe as we observe it, in which all distances appear to have expanded and to continue to expand at the same rate, separating matter without any apparent force. Gravitational attraction of matter may then simply be a reflection of the resistance of matter to this separation;

(12)   On the other hand, the universe may have begun literally with a big bang which resulted in a uniform increase in the distance between matter as it evolved;

(13)   The existence of a large amount of energy at the origin of the universe helps explain the subsequent expansion without the need to refer to it as dark energy;

(14)   The fact that matter, comprising protons, neutrons and electrons, were formed whilst the distance between them was increasing may have something to do with the emergence of gravity;

(15)   If the inward attraction of matter is a reflection of the resistance to the outward expansion of the universe, it may be possible to relate the gravitational constant G to the current rate of expansion of the universe;

(16)   Gravity appears to be some sort of entanglement of the protons, neutrons and electrons comprising matter that has existed since soon after the Big Bang;

(17)   Gravity appears to be similar to (non-relativistic) quantum entanglement between two particles that interact and then separate in such a way that the quantum state of each particle cannot be described independently of the state of the others?;

86

(18)   The mass of the observable universe (including ordinary matter, the interstellar medium (ISM), and the intergalactic medium (IGM), but excluding dark matter and dark energy) is around 1.5×10^{53} kg;

(19)   The actual density of atoms in the universe is equivalent to roughly 1 proton per 4 cubic meters = $1.6726 \times 10^{-27}/4 = 4.18 \times 10^{-28}$ kg m^{-3};

(20)   This implies that the current volume of the observable universe is $(1.5 \times 10^{53})/(4.18 \times 10^{-28}) = 3.6 \times 10^{80}$ m^3, and its radius is 4.4×10^{26} m;

(21)   The rate of expansion of the universe is estimated to be 73.3 kilometers per second per megaparsec (1 megaparsec = 3,260,000 light years = 3.0857×10^{19} km) or for every km from the Earth $73.3/3.0857 \times 10^{19} = 2.375 \times 10^{-18}$ km/sec;

(22)   According to mass-energy equivalence, $E = mc^2$, the energy equivalent of one kilogram of mass is 8.99×10^{16} joules, so the energy equivalent of the mass of the observable universe is around $1.5 \times 10^{53} \times 8.99 \times 10^{16} = 1.35 \times 10^{70}$ joules;

(23)   The zero-energy universe hypothesis proposes that the total amount of energy in the universe is exactly zero: its amount of positive energy in the form of matter is exactly canceled out by its negative energy in the form of gravity;

(24)   Alternatively, a "closed" universe, where the density parameter $\Omega > 1$, and Ω is defined as the average matter density of the universe divided by a critical value of that density, in which positive energy dominates, will eventually collapse in a "Big Crunch"; while an "open" universe, where $\Omega < 1$, in which negative energy dominates, will either expand indefinitely or eventually disintegrate in a "Big Rip";

(25)   According to the "closed" universe model, the universe might have started, after the initial Big Bang expansion, as a large sphere in space containing uniformly distributed matter, largely in the form of atoms and molecules, comprised of protons, neutrons and electrons, which continued to expand after the force causing the initial expansion ceased, based on the outward momentum of the matter;

(26)   However, there appears to be a problem. Based on the actual density of atoms in the universe, the time taken for molecules of hydrogen or cosmic dust to accrete due to the force of gravitation is far too long;

(27)   The existence of a very large amount of energy at the time of the origin of the current universe, makes the idea of a universe in which gravitational attractive forces eventually overcome the forces causing the expansion of the universe

particularly attractive, in that it provides an explanation for this energy and for the Big Bang without invoking dark matter or dark energy;

(28) Under this theory, the current universe, which originated about 13.8 billion years ago, evolved for about 9.6 billion years before a primitive form of life originated under the particular conditions of a small rocky planet;

(29) Gravitational energy or gravitational potential energy U is the potential energy a massive object m has in relation to another massive object M due to gravity; $U = GmM/R$, where R is the distance between the centers.   It is the potential energy associated with the gravitational field, which is released (converted into kinetic energy) when the objects fall towards each other. Gravitational potential energy increases when two objects are brought further apart.

In the common situation where a much smaller mass m is moving near the surface of a much larger object with mass M, the gravitational field is nearly constant and so the expression for gravitational energy can be simplified. The change in potential energy moving from the surface (a distance R from the center) to a height h above the surface is

$$\Delta U \approx GmM/R^2 \approx m(GM/R^2)\,h.$$

As the gravitational field is $g = GM/R^2$, this reduces to

$$\Delta U \approx mgh.$$

Taking $U = 0$ at the surface (instead of at infinity), the familiar expression for gravitational potential energy emerges:

$$U = mgh.$$

(30) This returns us to the question of whether it may be possible to relate the gravitational constant G to the current rate of expansion of the universe.

The gravitational field at the at the surface of the universe, is approximately equal to the gravitational constant. The gravitational field $g = GM/R^2$. Substituting $M = 1.5 \times 10^{53}$ kg and $R = 4.4 \times 10^{26}$ m, gives

$$g = 0.7748\ G.$$

$g = G$ when $R^2 = M = 1.5 \times 10^{53}$; or $R = 3.873 \times 10^{26}$ m. This implies that the volume of the universe, $V = 4/3 \, \pi R^3 = 2.4335 \times 10^{80}$ m³; which is in line with other estimates.

This is probably the closest that I can get.

6   Electricity & Magnetism   May 20, 2024.

The Heaviside formulation of Maxwell's Equations (1884).

$\nabla \cdot \mathbf{E} = \rho/\varepsilon_0 = 0$ Gauss's Law

$\nabla \times \mathbf{E} = - \partial\mathbf{B}/\partial t$ The Maxwell–Faraday version
                                                        of Faraday's Law of Induction

$\nabla \cdot \mathbf{B} = 0$ Gauss's Law for Magnetism

$\nabla \times \mathbf{B} = - \mu_0 (\mathbf{J} + \varepsilon_0\, \partial\mathbf{E}/\partial t) = - \mu_0\varepsilon_0\, \partial\mathbf{E}/\partial t$ The Maxwell-Ampère Law with
                                                        Maxwell's addition.

This is the extraordinary story of the development of our understanding of electricity and magnetism; from Coulomb's measurement of the force between two electric charges using a torsion balance in 1785; Volta's construction in 1799 of the first *electrochemical battery*, known as a *voltaic pile*; the discovery of the link between electricity and magnetism by Ørsted in 1819; their combination in the electromagnetic wave theory of light in Maxwell's extraordinary papers, between 1855 and 1865; the recognition that an electric current comprises a flow of free electrons based on Bohr's 1913 theory of the atom; and finally Heisenberg's brilliant demonstration in 1928 that ferromagnetism is a *quantum entanglement* effect resulting from the spin of the electron and the structure of the molecular lattice. Oliver Heaviside's story is a gem.

Prior to Coulomb's measurement of the electrostatic force between two charges little was known about *electricity* apart from the observations of shocks from electric fish, the creation of an electric spark by rubbing amber, and Benjamin Franklin's discovery in 1752 that lightning was a manifestation of electricity, and his explanation for the Leyden jar as a device for storing large amounts of electrical charge. Prior to Ørsted's discovery by the twitching of a compass needle near a wire, that an electric current could create a magnetic field, little was known about *magnetism*, apart from the fact that lodestones, naturally magnetized pieces of the mineral magnetite, could attract iron, the use of a lodestone compass for navigation in the 12th century, and William Gilbert's recognition in 1600 that the Earth was magnetic and that this was the reason compasses pointed north.

Part I describes what is general understood by electricity and magnetism. **Part II** provides a record of the discoveries and the resulting theories prior to Maxwell's papers on electromagnetism. **Part III** addresses electromagnetic radiation. **Part IV** reviews the explanation of electrical and magnetic phenomena in terms of the motion and spin of charge particles and quantum theory.

In **Part II**, Coulomb's paper, [Coulomb, C-A. (1785). *Premier mémoire sur l'électricité et le magnétisme.*], describes Coulomb's construction of a torsion balance and how he used this to demonstrate what he called the fundamental law of electricity, now known as *Coulomb's Law*. The law states that *the magnitude, or absolute value, of the attractive or repulsive electrostatic force between two point-charges is directly proportional to the product of the magnitudes of their charges and inversely proportional to the squared distance between them.*

In Volta, A.G.A.A. (1800). XVII. *On the Electricity excited by the mere Contact of conducting Substances of different kinds*, Volta describes his discovery that dissimilar metals separated by an electrolyte could create an electric current, and his construction of the first *electrochemical battery*, known as a *voltaic pile*, consisting of an alternating series of copper and zinc discs separated by brine-soaked cardboard. Within 20 years, this led to a cascade of discoveries of phenomena, theories and inventions relating to electricity and magnetism.

Poisson's paper, [Poisson, S.D. (1813). *Remarques sur une équation qui se présente dans la théorie des attractions des sphéroïdes.*], describes what became known as Poisson's Equation, a generalization of Laplace's equation, which is a partial differential equation of broad utility in theoretical physics. *The solution to Poisson's Equation is the potential field caused by a given electric charge or mass density distribution; with the potential field known, one can then calculate electrostatic or gravitational (force) field.*

Gauss's paper, [Gauss, C.F. (1813). *Theoria attractionis corporum sphaeroidicorum ellipticorum homogeneorum, methodo nova tractate.*], describes the electrical attraction between two ellipsoids, which became known as *Gauss's Law*, despite Lagrange's priority. *Gauss's Law states that the net electric flux through any hypothetical closed surface is equal to $1/\varepsilon_0$ times the net electric charge within that closed surface, where ε_0 is the permittivity of a vacuum* (also known as the *electric constant*). The *differential form* of Gauss's Law in Heaviside's vector notation is $\nabla \cdot \mathbf{E} = \rho/\varepsilon_0$ where $\nabla \cdot \mathbf{E}$ is the divergence of the *electric field*, ε_0 is the *permittivity of a vacuum* (also known as the *electric constant*), and ρ is the *volume charge density* (charge per unit volume). *Gauss's law*, and *Gauss's law for magnetism*, $\nabla \cdot \mathbf{B} = 0$, are two of the four Maxwell's equations.

Ørsted's paper, [Ørsted, H.C. (1820). *Experiments on the Effect of a Current of Electricity on the Magnetic Needle.*], describes his discovery that a compass needle is deflected from magnetic north by a nearby electric current, confirming a *direct relationship between electricity and magnetism.*

Biot, J.B. & Savart, F. (1820). *Note sur le magnétisme de la pile de Volta.* describes the experiments which led to what is now known as the Biot–Savart Law governing *the magnetic field generated by a constant electric current*. It relates the magnetic field to the magnitude, direction, length, and proximity of the electric current. *The Biot–Savart Law states that the magnetic flux density at a position in 3-dimensional space generated by a steady filamentary current around a closed path is proportional to the line integral around the curve of the current times the displacement vector of the wire element divided by the cube of the magnitude of the displacement vector.*

In Ampère, A-M. (1822). *Memoire sur la Determination de la formule qui represente l'action mutuelle de deux portions infiniment petites de conducteurs voltaiques.*, Ampère derives his force law, $F_m = 2k_A I_1 I_2 L/r$ where k_A is the *magnetic force constant* from the *Biot–Savart law*, F_m is the *total force* on either wire (the longer is approximated as infinitely long relative to the shorter), L is the length of the shorter wire, r is the distance between the two wires, and I_1, I_2 are the *direct currents* carried by the wires.

Apart from the 12-page introduction and a 1-page note, most of the 146-page memoire Ampère, A-M. (1826). *Théorie des phénomènes électro-dynamiques, uniquement déduite de l'expérience.* is devoted to describing experiments addressing different shapes, sizes and positions of the conductors.

Faraday, M.V. (1831). *Experimental Researches in Electricity. § 1. On the Induction of Electric Currents* describes the experiments from which Faraday derived his *Law of Induction*, which determines how an electric current produces a magnetic field and, conversely, how a changing magnetic field generates an electric current in a conductor. *Faraday's law states that the work done on a unit charge when it has travelled around a closed path (referred to as the electromotive force) is equal to the negative of the time rate of change of the magnetic flux enclosed by the path.* The *integral form* of Faraday's Law is $\int_{\Sigma(t)} \mathbf{E}(t) \cdot d\boldsymbol{l} = - \iint_{\Sigma(t)} \mathbf{B}(t) \cdot d\mathbf{A}$, where $\mathbf{E}(t)$ is the electric field, $\mathbf{B}$ is the *magnetic field*, $d\boldsymbol{l}$ is an infinitesimal vector element of the contour $d\Sigma$, $d\mathbf{A}$ is an infinitesimal vector element of surface $\Sigma(t)$, and $\mathbf{B} \cdot d\mathbf{A}$ is the *vector dot product* representing the element of flux through $d\mathbf{A}$. The *differential form* of Faraday's Law in Heaviside's vector notation is $\nabla \times \mathbf{E} = - \partial\mathbf{B}/\partial t$, where $\nabla \times$ is the *curl*.

In Faraday, M.V. (1845). *On the magnetization of light and the illumination of magnetic lines of force.*, Faraday introduced what became known as Faraday's *"lines of force"*, which was the driving force behind Maxwell's subsequent work on the electromagnetic radiation.

Part III, addresses *electromagnetic radiation*. It starts with a letter, Faraday, M.V. (1846). *Thoughts on Ray Vibrations.*, describing Faraday's speculation at the end of a lecture in

1846, that light could be a vibration of the *electric* and *magnetic lines of force*; an idea that led to Maxwell's electromagnetic wave theory of light.

Maxwell, J.C. (1855). *On Faraday's lines of force.*, written when Maxwell was 24 years old, is the first of Maxwell's three extraordinarily papers on electromagnetic radiation. Maxwell starts by noting that in order to make further progress with the science of electricity, the student must make himself familiar with a considerable body of most intricate mathematics, the mere retention of which in the memory materially interferes with further progress. He then explains the value of physical analogies in order to obtain physical ideas. He describes how this had been applied to the changes of direction which light undergoes in passing from one medium to another, which are identical with the deviations of the path of a particle in moving through a narrow space in which intense forces act. He notes that this analogy extended only to the direction and not to the velocity of motion, of light. He then refers to t*he analogy between light and the vibrations of an elastic medium, noting that it is founded only on a resemblance in form between the laws of light and those of vibrations.* He explains that i*t is by the use of analogies of this kind that I have attempted to bring before the mind those mathematical ideas which are necessary to the study of the phenomena of electricity*; attributing the methods to those suggested by the processes of reasoning which are found in the researches of Faraday.

He then proceeds to apply this to explain and illustrate Faraday's idea of *"lines of force"* using a *hydrodynamical* analogy. In order to provide some method of indicating the intensity of the force at any point he considers these c*urves not as mere lines, but as fine tubes of variable section carrying an incompressible fluid,* then, "since the velocity of the fluid is inversely as the section of the tube, we may make the velocity vary according to any given law, by regulating the section of the tube, and in this way, we might represent the intensity of the force as well as its direction by the motion of the fluid in these tubes". He then describes a method by which the consequences of assuming certain conditions of motion can be traced, and to show how the laws of the attractions and inductive actions of magnets and currents may be clearly conceived, without making any assumptions as to the physical nature of electricity.

In Maxwell, J.C. (1861 & 1862) *On Physical Lines of Force.*, a four-part paper, Maxwell uses his hydrodynamic analogy in conjunction with a "sea" of "molecular vortices" to model Faraday's lines of force. This paper is considered one of the most historically significant publications in physics and science in general, comparable with Newton's *Principia Mathematica.* In Part I of this paper Maxwell develops his hydrodynamic model. In Part II, he applies his hydrodynamic model to electric currents. In Part III, Maxwell applies his hydrodynamic model to the *wave motion of light*, and using the density and transverse elasticity of his sea of molecular vortices in *Newton's equation for the speed of sound*, obtains a value very close to the *speed of light*. Maxwell concludes that "we can scarcely avoid the inference that *light consists in the transverse undulations of the same medium which is the cause of electric and magnetic phenomena*".

In Maxwell, J.C. (1865). *A Dynamical Theory of the Electromagnetic Field.*, Maxwell notes that the most obvious mechanical phenomenon in electrical and magnetical experiments is *the mutual action by which bodies in certain states set each other in motion while still at a sensible distance from each other*. He adds that the first step in reducing these phenomena into scientific form was to ascertain the *magnitude and direction of the force* acting between the bodies, and when it was found that this force depended upon the relative position of the bodies and on their electric or magnetic condition, it seemed at first sight natural to explain the facts by assuming the existence of something either at rest or in motion in each body, constituting its electric or magnetic state, and capable of acting at a distance according to mathematical laws. In this way mathematical theories of statical electricity, of magnetism, of the mechanical action between conductors carrying currents, and of the induction of currents had been formed. In these theories the force acting between the two bodies was treated with reference only to the condition of the bodies and their relative position, and without any express consideration of the surrounding medium. These theories assumed, more or less explicitly, the existence of substances the particles of which have the property of acting on one another at a distance by attraction or repulsion.

But Maxwell then notes that, in view of the difficulties that he perceived with this approach, he had preferred to seek an explanation in another direction, *by supposing the force to be produced by actions which go on in the surrounding medium as well as in the excited bodies*, and endeavoring to explain the action between distant bodies without assuming the existence of forces capable of acting directly at sensible distances. Maxwell describes his theory as a *theory of the Electromagnetic Field*, because it had to do with the space in the neighborhood of the electric or magnetic bodies, and a *Dynamical Theory*, because it assumed that in that space there is matter in motion, by which the observed *electromagnetic* phenomena were produced. He introduces the role of the *electromotive force* (defined by Faraday as the work done on a unit charge when it has travelled around a closed path); *when this electromotive force acts on a conducting circuit, it produces a current*; *when it acts on a dielectric it produces an electric displacement*. He notes that we know that when an electric current is established in a conducting circuit, the neighboring part of the field is characterized by certain magnetic properties, and that if two circuits are in the field, the magnetic properties of the field due to the two currents are combined. Thus, *each part of the field is in connection with both currents, and the two currents are put in connection with each other in virtue of their connection with the magnetization of the field.*

94

Maxwell observes that the first result of this connection that he proposed to examine, was the *induction* of one current by another, and by the motion of conductors in the field. The second result, which is deduced from this, was the mechanical action between conductors carrying currents. He then applies the phenomena of *induction and attraction of currents to the exploration of the electromagnetic field*, and the laying down systems of *lines of magnetic force* which indicated its magnetic properties. In order to bring these results within the power of symbolical calculation, he then expresses them in the form of his *General Equations of the Electromagnetic Field*.

In part III of the paper, Maxwell formulates twenty equations which were to become known as *Maxwell's equations*, until this term became applied instead to a vectorized set of four equations formulated by Oliver Heaviside in 1884, which had all appeared in his 1861 paper *On Physical Lines of Force*. The Heaviside formulation of Maxwell's Equations (1884) comprises 4 equations derived from 12 out of the 20 of Maxwell original equations:

$\nabla \cdot \mathbf{E} = \rho/\varepsilon_0 = 0$ (Gauss's law);

$\nabla \times \mathbf{E} = -\partial\mathbf{B}/\partial t$ (The Maxwell–Faraday version of Faraday's law of induction);

$\nabla \cdot \mathbf{B} = 0$ (Gauss's law for magnetism)

$\nabla \times \mathbf{B} = -\mu_0 (\mathbf{J} + \varepsilon_0 \partial\mathbf{E}/\partial t) = -\mu_0\varepsilon_0 \partial\mathbf{E}/\partial t$ (Ampère's law with Maxwell's addition).

In Part IV, Maxwell notes that in the first part of the paper he had applied the *optical hypothesis* of an elastic medium through which the vibrations of light are propagated, in order to show that he had warrantable grounds for seeking, in the same medium, the cause of other phenomena as well as those of light. He had then examined *electromagnetic phenomena*, seeking for their explanation in the properties of the field which surrounds the electrified or magnetic bodies. In this way he had arrived at *equations expressing properties of the electromagnetic field*. He now proceeds to investigate whether these properties of the *electromagnetic field*, deduced from electromagnetic phenomena alone, were sufficient to explain the propagation of light through the same substance. By these means he obtains an equation relating the number of electrostatic units in one electromagnetic unit to the *speed of light*, and found that this agreed with the velocity of light deduced from experiment.

The Heaviside formulation of Maxwell's Equations is then reviewed. In 1884 Oliver Heaviside recast Maxwell's mathematical analysis from its original cumbersome form to its modern vector terminology, thereby *reducing twelve of the original twenty equations in twenty unknowns down to the four differential equations in two unknowns we now know as Maxwell's equations*. The four re-formulated Maxwell's equations describe the nature of *electric charges* (both static and moving), *magnetic fields*, and the relationship between the two, namely *electromagnetic fields*. Heaviside presented these equations *in modern vector format* using the *nabla operator* (∇) devised by William Rowan Hamilton in 1837.

Heaviside *eliminated the vector and scalar potentials A and Ψ from the equations and expressed the electromagnetic relations purely in terms of the electric and magnetic fields, E and H*. His most important step was to derive what he called the *second circuital law*, which relates the curl of **E** to the rate of change of **H**. Starting with Maxwell's relation **E** $= -\partial\mathbf{A}/\partial t - \nabla\Psi$, Heaviside took the curl of both sides: $\nabla \times \mathbf{E} = \nabla(-\partial\mathbf{A}/\partial t) - \nabla \times \nabla\Psi$. Since the curl of any gradient is zero, the last term vanishes. By switching the order of the space and time differentiations, Heaviside obtained $\nabla \times \mathbf{E} = -\partial(\nabla \times \mathbf{A})/\partial t$. Since $\nabla \times \mathbf{A} = \mu\mathbf{H}$, this yielded the *second circuital law*, $\nabla \times \mathbf{E} = -\mu\,\partial\mathbf{H}/\partial t$. Heaviside then combined this with equations drawn from Maxwell's *Treatise* to obtain his new set of four "Maxwell's equations":

$$\nabla \cdot \varepsilon\mathbf{E} = \rho$$
$$\nabla \times \mathbf{E} = -\mu\,\partial\mathbf{H}/\partial t$$
$$\nabla \cdot \mu\mathbf{H} = 0$$
$$\nabla \times \mathbf{H} = k\mathbf{E} + \varepsilon\,\partial\mathbf{E}/\partial t,$$

where ε is the *permittivity*; μ the *permeability*; ρ the *charge density*; and k the *conductivity*.

Part IV reviews the explanation of electrical and magnetic phenomena in terms of the motion and spin of charge particles and quantum theory. It was not until the elucidation of Bohr's model of the atom in 1913 that it was understood that both were manifestations of the movement of fundamental particles of matter with an electric charge, in particular the electron and the proton, and not until Heisenberg's paper in 1928 that it was understood that ferromagnetism was a quantum phenomenon, resulting from the quantum entanglement of the electron's spin.

In Part I of Bohr, N. (1913). *On the Constitution of Atoms and Molecules.*, Bohr adapts Rutherford's theory of atomic structure to Planck's quantum hypothesis to create his model of the atom. He supposes the atom to consist of a nucleus with a positive charge Ze, and Z electrons with charge – e each, moving according to the laws of classical mechanics. He introduces the idea that an electron could drop from a higher-energy orbit to a lower one, in the process emitting a quantum of discrete energy. This became a basis for what is now known as the *old quantum theory*. From a set of assumptions concerning the stationary state of an atom and the frequency of the radiation emitted or absorbed when the atom passes from one such state to another, he shows that it is possible to obtain a simple interpretation of the main laws governing the line spectra of the elements, and to deduce the Balmer formula for the hydrogen spectrum. He addresses the mechanism of the binding of electrons by a positive nucleus in relation to Planck's theory. Part II of this paper, [Bohr, N. (1913). *On the Constitution of Atoms and Molecules. Part II. Systems Containing only a Single Nucleus.*] applies Bohr's model of the atom to systems containing a single nucleus.

Compton's paper, [Compton, A.H. (1921). *The Magnetic Electron.*], on investigations of ferromagnetic substances with X-rays, was the first to introduce the idea of *electron spin*. Compton hypothesizes that the electron *magnetic moment* is intrinsically connected to the electron's *spin*.

Without being aware of Compton's suggestion Uhlenbeck and Goudsmit note in their paper, [Uhlenbeck, G.E. & Goudsmit, S. (1925). *Ersetzung der Hypothese vom unmechanischen Zwang durch eine Forderung bezuglich des inneren Verhaltens jedes einzelnen Elektrons.* (Replacement of the hypothesis of unmechanical coercion by a requirement regarding the internal behavior of each individual electron.)], that doublets in the alkali spectra did not conform to current models of the atom. They propose applying the model of the *spinning electron* to interpret a number of features of the quantum theory of the *anomalous Zeeman effect*, and apply the classical formula for spherical rotating electron with finite radius and surface charge.

Heitler, W. & London, F. (1927). *Wechselwirkung neutraler Atome und homöopolare Bindung nach der Quantenmechanik.* (Interaction of neutral atoms and homeopolar bonding according to quantum mechanics.) examines the interaction between *neutral atoms* though non-polar bonds, in what is known as valance bonds. Quantum mechanics is applied to the calculation of the *interaction energy* of the atoms when they move closer together. Due to *quantum entanglement*, it is found that two neutral atoms can interact with each other in two ways; *the problem is twofold degenerate, corresponding to the two ways of assigning the electrons to the neutral atoms.* Examination of the different cases of two H atoms and two He atoms shows that by applying the *Pauli principle*, the selected eigenfunctions of the system change or maintain their sign respectively when two electrons are swapped if the two electrons compared have the same or different *spin*. It is found that in the case of He there is only one solution, which yields about the right size of the He gas kinetic-radius, *due to the fact that 2 He atoms (and the same applies to all noble gases) cannot differ in their spin* – in contrast to hydrogen (and all atoms with unfinished shells) – so that 2 He atoms have only one possible mode of behaving.

In Heisenberg, W. (1928). *Zur Theory of Ferromagnetismus.*, another brilliant paper, Heisenberg notes that empirical results exhibited *ferromagnetism* as an entirely similar state of affairs to what was previously observed in the spectrum of the helium atom. It seems to follow from the levels in the helium atoms that a powerful interaction prevails between the spin directions of two electrons that led to the splitting of the level structure into systems of singlets and triplets. He also notes that this is closely related to explaining ferromagnetic phenomena as being implied by the *exchange phenomenon* (resulting from *quantum entanglement*). He proceeds to show that the *Coulomb interaction*, together with the *Pauli principle*, succeeds in evoking the same effects as the molecular field that Weiss had postulated, noting that it was only in recent times that mathematical methods were developed for the treatment of such a complicated problem in the important investigations of Wigner, Hund, and Heitler & London. Heisenberg assumes as a first approximation that the lattice separations are very large, and that every electron belongs to its own atom, and applies Heitler-London's calculations to the case of 2n electrons in a state of interaction, finding 2n electrons in 2n positionally different quantum cells. Due to their smallness, he was able to leave the magnetic interactions outside of consideration, and shows that *the spin moments of all electrons become partly parallel and partly anti-parallel as a result of the exchange processes*. By adding the fundamental Pauli principle to this, viz., that the eigenfunctions of the total system should be *anti-symmetric* in all electrons, he shows that an entirely well-defined *total magnetic moment* belongs to each level value of the perturbed system, and there are (2n)! levels in the unperturbed system (ignoring the Pauli principle and spin). He then shows that a *statistical treatment of ferromagnetism is possible when all energy values had been calculated.* Heisenberg concludes that *an atom in a lattice can only be exchanged with its "neighbors"*; exchanges with atoms that lie further away that the "neighboring atoms" can be neglected. The *number of "neighbors" of an atom* is, e.g., 1 in a molecular lattice of diatomic molecules, 2 in a linear chain, 4 in a quadratic surface lattice, 6 in a simple cubic lattice, 8 in a cubic, space-centered lattice, and 12 in a cubic, face-centered lattice. By assuming a distribution of energy values about the mean has the approximate form of a Gaussian error curve, Heisenberg shows that small or negative values of the constant β [$= zJ_0/kT$)] result in *paramagnetism*; and that *ferromagnetism is only possible for lattice types for which an atom has at least eight neighbors*, which is the case for Fe, Co, Ni, whose lattices are all cubic, some of which are space-centered ($z = 8$) and some of which are face-centered ($z = 12$). He concludes that two conditions are necessary for the appearance of *ferromagnetism*: the crystal lattice must be a type such that *any atom has at least 8 neighbors*; and the *principal quantum number* of the electrons that are responsible for magnetism must be $n \geq 3$.

In Dirac, P.A.M. (1929). *Quantum Mechanics of Many-Electron Systems.*, Dirac comments that the general theory of quantum mechanics is now almost complete, and that the imperfections that still remained were in connection with the exact fitting in of the theory with relativity ideas, which only gave rise to difficulties onwhen high-speed particles were involved and were therefore of no importance in the consideration of atomic and molecular structure and ordinary chemical reactions. The difficulty is only that the exact application of these laws leads to equations much too complicated to be soluble. He notes that it is desirable that approximate practical methods of applying quantum mechanics should be developed which can lead to an explanation of the main features of complex atomic systems without too much computation; current *non-relativistic* quantum theory cannot give an explanation of *multiplet structure* without an extraneous assumption of large forces coupling the *spin vectors* of the electrons in an atom. The explanation is provided by *quantum entanglement* through *exchange interaction* arising from electrons being indistinguishable one from another resulted in large *exchange energies* between electrons in different atoms. This accounted for homopolar valency bonds; for each *stationary state* of the atom there was one magnitude of the total spin vector. He also notes that developments of the *theory of exchange* made by Heitler & London and Heisenberg make extensive use of *group theory*, which is a theory of certain quantities that do not satisfy the commutative law of multiplication and should thus form a part of quantum mechanics, and then translates the methods and results of *group theory* into the language of *quantum mechanics*. He demonstrates that *exchange interaction* equal to a constant *perturbation energy*, together with *coupling energy* between spin vectors, determines energy levels; and shows that in the first approximation the *exchange interaction* between the electrons can be replaced *by a coupling between their spins*, the energy of this coupling for each pair of electrons being equal to the scalar product of their *spin vectors* multiplied by a numerical coefficient given by the *exchange energy*.

In Dirac, P.A.M. (1931). *Quantized singularities in the electromagnetic field.*, Dirac notes that the object of this paper is to show that quantum mechanics does not preclude the existence of *isolated magnetic poles*. He addresses the fact that the smallest electric charge known experimentally, e, is given by $hc/e^2 = 137$, and considers a particle whose motion is represented by a wave function. Dirac uses *non-relativistic* theory to show that the change in phase round a closed curve must be same for all wave functions, and applies this to the motion of an electron in an electromagnetic field. He shows that non-integrable derivatives of the phase of the wave function represent *potentials* of the electromagnetic field. This leads to wave equations whose only physical interpretation is in the motion of an electron in the field of a single pole. This does not give a value for e but shows *reciprocity between electricity and magnetism*, and that *the strength of pole and electric charge must both be quantized*. It gave the relationship between the strength of the quantum of magnetic pole and electronic charge $hc/e\mu_0 = 2$ but *did not explain their magnitudes*. Dirac suggests that the reason that isolated magnetic poles have not been separated is probably due to the very large force between two one-quantum poles of opposite sign, $(137/2)^2$ times that of that between electron and proton.

In my last book, "Electricity & Magnetism", two conclusions stood out; the *rotary* nature of magnetism compared with the *linear* translation of electricity; and the dependence of *ferromagnetism* on *entanglement* of *electrons* with the same *quantum spin states*. This book takes a closer look at *quantum entanglement*.

Entanglement occurs between *valence electrons* with the same *quantum spin states* creating *non-polar (valence) bonds.*

> [*Valence electrons* are electrons in the outermost shell of an atom, and that can participate in the formation of a chemical bond if the outermost shell is not closed.]

Entanglement also occurs between electrons with the *same quantum spin states* in *materials* in a *magnetic field*, creating various types of *magnetism*, depending on structure of the atomic lattice of the material. It has also been demonstrated that *quantum entanglement* occurs *at a distance* between electrons, photons, top quarks, and molecules.

Part I describes the development of the current theory of the *spin of the electron*. **Part II** describes the development of the theory of how *entanglement* between electrons with the same *quantum spin states* result in *exchange interaction*. **Part III** addresses the quantum theory of the susceptibility of *materials* to a *magnetic field* and the resulting types of magnetism. **Part IV** describes *quantum entanglement at a distance* between electrons and other particles with the same *quantum spin states*.

Part I (Electron Spin) starts with Arthur Compton's 1921 paper, [Compton, A.H. (1921). *The Magnetic Electron.*], in which the idea of a *quantized spinning of the electron* was put forward for the first time. Compton hypothesized that the electron *magnetic moment* was intrinsically connected to the electron's *spin*.

This is followed by Wolfgang Pauli,'s 1925 paper, [Pauli, W. (1925). *Über den Zusammenhang des Abschlusses der Elektronengruppen im Atom mit der Komplexstruktur der Spektren.* (On the connection between the closure of the electron groups in the atom and the complex structure of the spectra.] in which Pauli first reviewed the established theories for the energy differences *of the triplet levels of the alkaline earths*, based respectively, on *the anomaly of the relativity correction* of the *optically active electron*, and *the dependence of the interaction between the electron and the atom core on the relative orientation of these two systems*. He noted a serious difficulty with the former is the connection of these ideas with the *correspondence principle*, which was well known to be a necessary means to explain the selection rules for the *quantum numbers* k_1, j, and m and the polarization of the Zeeman components, in particular that *it was necessary that the totality of the stationary states of an atom corresponded to a collection (class) of orbits with a definite type of periodicity properties.* The dynamic explanation of this kind of motion of the *optically active electron,* which was based upon the assumption of deviations of the forces between the *atom* core and the *electron* from central symmetry, *seemed to be incompatible with the possibility to represent the alkali doublet (and thus also the magnitude of the corresponding precession frequency) by relativistic formulae.* Consequently, Pauli, decided to apply instead the alternative *non-relativistic* theory to the problem of *completion of electron groups in an atom,* in order to draw conclusions about the *number of possible stationary states* of an *atom* when several equivalent *electrons* are present. But this did not address the position and relative order of the term values. On the basis of these results, Pauli obtained a general classification of every *electron* in the *atom* by the principal quantum number n and two auxiliary quantum numbers k_1 and k_2 to which he added a further quantum number m_1 in the presence of an external field, in agreement with experiments. In particular, his rule explained Stoner's result in a natural way and with it the period lengths 2, 8 18, 32.

Then, as noted above, without being aware of Compton's suggestion, Uhlenbeck and Goudsmit noted in their paper, [Uhlenbeck, G.E. & Goudsmit, S. (1925). *Ersetzung der Hypothese vom unmechanischen Zwang durch eine Forderung bezuglich des inneren Verhaltens jedes einzelnen Elektrons.*], that doublets in the alkali spectra did not conform to current models of the atom. They applied the model of the *spinning electron* to interpret a number of features of the quantum theory of the *anomalous Zeeman effect,* using the classical formula for spherical rotating electron with finite radius and surface charge.

This is followed by Paul Dirac's 1928 paper, [*The Quantum Theory of the Electron.*], in which Dirac noted that the new quantum mechanics applied to the problem of the structure of the atom with *point-charge electrons* did not give results in agreement with experiment. The discrepancies consisted of "duplexity" phenomena; the observed number of stationary states for an electron in an atom being twice the number given by the theory. Goudsmit and Uhlenbeck had introduced the idea of an electron with a *spin.* Previous *relativity* treatments by Gordon and Klein obtained the operator of the wave equation by the same procedure as in the *non-relativity* theory; they substituted classical *quantum differential operators* for the *momentum vector* in the amended *relativistic Hamiltonian equation* and applied the resulting differential operator to the *wave function* to obtain the *Klein-Gordon equation.* Dirac noted that Gordon and Klein's treatments gave rise to two difficulties. The *first difficulty* was in the physical interpretation of wave-mechanical expressions for the *charge* and the *current.* This was satisfactory for emission and absorption of radiation, but only provided the probability of any dynamical variable at any specific time having a value between specified limits if they referred to the *position* of the electron, but, unlike the *non-relativity* theory, *not if they refer to its momentum or any other dynamical variable.* The *second difficulty* was that the conjugate imaginary of the *wave equation* was the same as that for an electron with charge – e and negative energy.

This paper only addressed the removal of the first of difficulties. The resulting theory was only an approximation but appeared sufficient to address the duplexity problems without further assumptions. Dirac applied the method of *q-numbers* and using non-commutative algebra exhibited the properties of a free electron and of an electron in a central field of electric force. He showed that simplest Hamiltonian for a *point charge electron satisfying requirements of both relativity and the general transformation theory* of quantum mechanics led to an explanation of all duplexity phenomena of number of stationary states being twice the observed value *without further assumption about spin*. In contrast to the Schrödinger equation which described wave functions of only one complex value, Dirac introduced *vectors of four complex numbers* (known as bispinors). This resulted in a *relativistic equation of motion* for the *wave function of the electron* referred to as the *Dirac equation*, $\{p_0 + p_1 (\boldsymbol{\sigma}, \mathbf{p}) + p_3 mc\} \psi = 0$, where $\mathbf{p}$ is the *momentum* vector, and $\boldsymbol{\sigma}$ denotes the vector ($\sigma_1, \sigma_2, \sigma_3$). This included a term equal to the spin correction given by Darwin and Pauli. It described all spin-½ particles with mass, but did not address the second class of solutions of the wave equation in which *charge of the electron is positive* and *energy of a free electron is negative*.

In the second part of this paper [(1928) *The Quantum Theory of the Electron. Part II.*], Dirac applied the *Dirac equation* to the conservation theorem, the selection principle, the relative intensities of the lines of a multiplet, and the Zeeman effect.

The last paper in this section [Pauli, W. (1940). *The Connection Between Spin and Statistics.*], appeared to be a rather peculiar effort to reinstate Einstein's *theory of special relativity*, possibly due to the fact that it was published at the outbreak of World War II, just prior to Pauli and Einstein both moving to Princeton. Pauli demonstrated that for a *relativistically invariant wave equation* for free particles: from postulate (I), *according to which the energy must be positive, the necessity of Fermi-Dirac statistics for particles with arbitrary half-integral spin*; from postulate (II), *according to which observables on different space-time points with a space-like distance are commutable, the necessity of Einstein-Bose statistics for particles with arbitrary integral spin*. Postulate I was introduced because "in case of *half-integral spin, …, a positive definite energy density, as well as a positive definite total energy, is impossible*". Similarly, postulate II was introduced so that "all physical quantities at finite distances exterior to the light cone … are commutable. … The justification for our postulate lies in the fact that measurements at two space points with a space-like distance can never disturb each other, *since no signals can be transmitted with velocities greater than that of light*".

In the first paper in **Part II** (Exchange Interaction), [Dirac, P. A. M. (October, 1926). *On the Theory of Quantum Mechanics.*], Dirac developed a *relativistic* treatment of *Schrodinger's wave theory* from a more general point of view in which the time t and its conjugate momentum – W were treated from the beginning on the same footing as the other variables. He applied his *relativistic formulation* to a system containing an atom with two electrons and found that *if the positions of the two electrons were interchanged the new state of the atom was physically indistinguishable from the original one.* In order that that the theory only enabled calculation of *observable quantities* it was necessary to treat (*mn*) and (*nm*) as only one *state*. *Unsymmetrical* functions of the co-ordinates (and momenta) of the two electrons could not be represented by matrices. *Symmetrical* functions such as the total *polarizations* of the atom could be considered to be represented by matrices without inconsistency. These matrices were by themselves sufficient to determine all the physical properties of the system. The *theory of uniformizing variables* introduced by the author *could no longer apply.* The new theory allowed two solutions satisfying the necessary conditions; one led to Pauli's principle that not more than one electron can be in any given orbit, and the other, when applied to the analogous problem of the ideal gas, led to the Einstein-Bose statistical mechanics. *With neglect of relativity mechanics this accounted for the absorption and stimulated emission of radiation* and showed that the elements of the matrices representing the total polarization determined the transition probabilities. *This could not be applied to spontaneous emission.*

This is followed by the path-breaking paper [Heitler, W. & London, F. (1927). *Wechselwirkung neutraler Atome und homöopolare Bindung nach der Quantenmechanik.* (Interaction of neutral atoms and homeopolar bonding according to quantum mechanics.).], in which Walter Heitler and Fritz London examined the interaction between *neutral atoms* known as *non-polar valance bonds*. They applied quantum mechanics to the calculation of the *interaction energy* of the atoms when they move closer together. Due to *quantum entanglement*, it was found that two neutral atoms could interact with each other in two ways. *The problem was twofold degenerate, corresponding to the two ways of assigning the electrons to the neutral atoms.* Examination of the different cases of two H atoms and two He atoms showed that by applying the *Pauli principle*, the selected eigenfunctions of the system changed or maintained their sign, respectively, when two electrons were swapped if the two electrons compared had the same, or different, *spin*. It was found that in the case of He there was only one solution, which yields about the right size of the He gas kinetic-radius, *due to the fact that 2 He atoms (and the same applied to all noble gases) could not differ in their spin* – in contrast to hydrogen (and all atoms with unfinished shells) – so that 2 He atoms had only one possible mode of behaving.

Then, in another paper, [Dirac, P. A. M. (1929). *Quantum Mechanics of Many-Electron Systems.*], in which he introduced the term *exchange interaction*, Dirac noted that the general theory of quantum mechanics was now almost complete, and that the imperfections that still remained were in connection with the exact fitting in of the theory with *relativity ideas*. They only gave rise to difficulties when high-speed particles were involved and were therefore *of no importance in the consideration of atomic and molecular structure and ordinary chemical reactions*. The difficulty was only that the exact application of these laws led to equations much too complicated to be soluble. Dirac noted that it was desirable that approximate practical methods of applying quantum mechanics should be developed which could lead to an explanation of the main features of complex atomic systems without too much computation. Current *non-relativistic* quantum theory could not give an explanation of *multiplet structure* without an extraneous assumption of *large forces coupling the spin vectors of the electrons in an atom*. The explanation was provided by *exchange interaction* arising from electrons being indistinguishable one from another resulted in large *exchange energies* between electrons in different atoms. This had accounted for *homopolar valency bonds*, in which, for each *stationary state* of the atom there was one magnitude of the *total spin vector*. Dirac also noted that developments of the *theory of exchange* made by Heitler & London and Heisenberg made extensive use of *group theory*, which was a theory of certain quantities that did not satisfy the commutative law of multiplication and should thus form a part of quantum mechanics, and then translated the methods and results of *group theory* into the language of *quantum mechanics*. He demonstrated that *exchange interaction* equal to a constant *perturbation energy*, together with *coupling energy* between spin vectors, determined *energy* levels; and showed that in the first approximation the *exchange interaction* between the electrons could be replaced *by a coupling between their spins*, the energy of this coupling for each pair of electrons being equal to the scalar product of their *spin vectors* multiplied by a numerical coefficient given by the *exchange energy*.

PART III (Susceptibility of materials to a magnetic field) begins with another brilliant paper by Heisenberg [Heisenberg, W. (1928). *Zur Theory of Ferromagnetismus*. (On the theory of ferromagnetism.)], in which Heisenberg noted that empirical results exhibited *ferromagnetism* as an entirely similar state of affairs to what was previously observed in the spectrum of the helium atom. It seemed to follow from the levels in the helium atoms that *a powerful interaction prevailed between the spin directions of two electrons* that led to the splitting of the level structure into systems of singlets and triplets. He also noted that this was closely related to explaining ferromagnetic phenomena as being implied by the *exchange phenomenon* (resulting from *quantum entanglement*). Heisenberg proceeded to show that the *Coulomb interaction*, together with the *Pauli principle*, succeeded in evoking the same effects as the molecular field that Weiss had postulated, noting that it was only in recent times that mathematical methods were developed for the treatment of such a complicated problem in the important investigations of Wigner, Hund, and Heitler & London. Heisenberg assumed as a first approximation that the lattice separations were very large, and that every electron belonged to its own atom, and applied Heitler-London's calculations to the case of 2n electrons in a state of interaction, finding 2n electrons in 2n positionally different quantum cells. Due to their smallness, he was able to leave the magnetic interactions outside of consideration, and showed that *the spin moments of all electrons become partly parallel and partly anti-parallel as a result of the exchange processes*. By adding the fundamental Pauli principle to this, viz., that the eigenfunctions of the total system should be *anti-symmetric* in all electrons, he showed that an entirely well-defined *total magnetic moment* belonged to each level value of the perturbed system, and there were (2n)! levels in the unperturbed system.

> [An *eigenfunction* of a *linear operator* defined on some *function space* is any non-zero function in that space that, when acted upon by the linear operator, is only multiplied by some scaling factor called an *eigenvalue*. In quantum mechanics, the Schrödinger equation with the Hamiltonian operator can be solved by separation of variables if the Hamiltonian does not depend explicitly on time. In that case, the *wave function* leads to two differential equations, which are *eigenvalue* equations. The *eigenfunctions* of the Hamiltonian operator are *stationary states* of the quantum mechanical system, each with a corresponding *energy*. They represent allowable *energy states* of the system and may be constrained by boundary conditions.]

He then showed that a *statistical treatment of ferromagnetism was possible when all energy values had been calculated*. Heisenberg concluded that *an atom in a lattice could only be exchanged with its "neighbors"*; exchanges with atoms that lie further away that the "neighboring atoms" could be neglected. The *number of "neighbors" of an atom* was, e.g., 1 in a molecular lattice of diatomic molecules, 2 in a linear chain, 4 in a quadratic surface lattice, 6 in a simple cubic lattice, 8 in a cubic, space-centered lattice, and 12 in a cubic, face-centered lattice. By assuming a distribution of energy values about the mean had the approximate form of a Gaussian error curve, Heisenberg showed that small or negative values of the constant β [$= zJ_0/kT$)] result in *paramagnetism*; and that *ferromagnetism is only possible for lattice types for which an atom had at least eight neighbors*, which was the case for Fe, Co, Ni, whose lattices are all cubic, some of which were space-centered ($z = 8$) and some of which were face-centered ($z = 12$). He concluded that two conditions were necessary for the appearance of *ferromagnetism*: the crystal lattice must be a type such that *any atom had at least 8 neighbors*; and the *principal quantum number* of the electrons that were responsible for magnetism must be $n \geq 3$.

Extracts are then provided from John Van Vleck's 400-page book [Van Vleck, J. H. (1932). *The Theory of Electric and Magnetic Susceptibilities.*], which provides an extremely comprehensive account of the *non-relativistic* quantum theory of the *electric and magnetic susceptibilities of materials*, including the quantum theory of *diamagnetism* and *paramagnetism* and detailed explanations of the *exchange effect* and Heisenberg's *theory of ferromagnetism*. Van Vleck noted in the Preface, "the analysis of experimental *magnetic susceptibilities* cannot be attempted until the quantum chapters, since *the numerical values of magnetic susceptibilities are inextricably connected with the quantization of angular momentum*". He also noted that *the theory of the electron spin may be presented in two ways,* viz. by means of what he calls a *semi-mechanical (Uhlenbeck-Goudsmit) model)* or by means of *Dirac's 'quantum theory of the electron'*. In the *semi-mechanical model*, matrix expressions for the *spin angular momentum* are written down by analogy with the *orbital angular momentum* matrices, with certain postulates regarding the occurrence of a half-quantum of *spin* per electron and *the ratio of spin magnetic moment to spin angular momentum* in order to bring the theory into line with experiments. He observed that the interaction of the *spin* with *external magnetic fields* was handled perfectly well by the *semi-mechanical model*, but Einstein's *special theory of relativity* was inconsistent with it. However, *the extension of Dirac's relativistic theory to many electron systems was at that time in a rather unsettled state* resulting in a nonvanishing probability of the mass of the electron changing sign, *an obvious absurdity.* So Van Vleck decided to present the quantitative aspects of the spin entirely with the aid of the older *non-relativistic semi-mechanical model*. In considering *diamagnetism*, [the property of materials that are repelled by a magnetic field; which creates an induced magnetic field in them in the opposite direction] he supposed that the atoms were in *singlet S states* [in which all electrons in the S orbital are paired], as otherwise there was an overwhelming *paramagnetism* [the form of magnetism whereby some materials are weakly attracted by an externally applied magnetic field, and form internal induced magnetic fields in the direction of the applied magnetic field]. In such *states* there cannot be even an instantaneous *magnetic moment* in the absence of external fields. Van Vleck distinguished other cases, including where the *inter-atomic forces* were so small that the magnetism could be calculated by treating the atoms of the solid to be free (exemplified by *rare earth salts*); where *the orbital and spin magnetic effects were both largely destroyed*, resulting in feeble *paramagnetism* (which most elements exhibit in the solid state); and solids in which *the Heisenberg exchange forces tended to align the spins parallel and so create ferromagnetism* (as for iron, nickel, and cobalt). Van Vleck noted that the *exchange forces had the effect of introducing a very strong coupling between the spins of paramagnetic atoms or ions. Diamagnetic* atoms or ions have no resultant spin and so do not give rise to any *exchange forces* tending to orient the *spins* of other atoms. Moreover, these *exchange forces* became of subordinate importance where the density of *paramagnetic* atoms or ions was low, because the great majority of the atoms are *diamagnetic*, and in most salts involving the iron group, which are consequently only *paramagnetic*. He then described how *entanglement* between electrons with the same *quantum spin states* creates an *exchange effect* which results in a strong coupling between their *spins*, and concluded by explaining how Heisenberg applied this in his theory of *ferromagnetism*.

Part IV. (Quantum Entanglement) addresses *quantum entanglement* at a distance, which Einstein referred to as "spooky action at a distance." The first paper in this section is what is known as the EPR paper, [Einstein, A., Podolsky, B. & Rosen, N. (1935). *Can Quantum-Mechanical Description of Physical Reality Be Considered Complete?*], in which the authors argued that "in a complete theory there is an element corresponding to each element of reality. *A sufficient condition for the reality of a physical quantity is the possibility of predicting it with certainty, without disturbing the system.* In quantum mechanics in the case of two physical quantities described by non-commuting operators, *the knowledge of one precludes the knowledge of the other.* Then either (1) the description of reality given by the *wave function* in quantum mechanics is not complete or (2) these two quantities cannot have simultaneous reality. Consideration of the problem of making predictions concerning a system on the basis of measurements made on another system that had previously interacted with it leads to the result that if (1) is false then (2) is also false. One is thus led to conclude that the description of reality as given by a *wave function* is not complete." Einstein was wrong, again.

This is followed by a brilliant rebuttal by Erwin Schrodinger [Schrodinger, E. (1935). *Discussion of probability relations between separated systems.*], in which he introduced the word *"entanglement"*. Schrodinger argued that when two systems, of which we know the states by their respective representatives, enter into temporary physical interaction due to known forces between them, and when the systems separate again, then they can no longer be described in the same way as before, viz. by endowing each of them with a representative of its own. *"I would not call that one but rather the characteristic trait of quantum mechanics*, the one that enforces its entire departure from classical lines of thought. *By the interaction the two representatives (or ψ-functions) have become entangled".*

30 years later, in 1964, John Stewart Bell, a physicist from Northern Ireland, determined that *quantum mechanics was incompatible with local hidden-variable theories* given some basic assumptions about the nature of measurement, subsequently known as *Bell's theorem.* [Bell, J.S. (1964). *On the Einstein-Podolsky-Rosen Paradox.*] "Local" here referred to the principle of *locality*, the idea that a particle could only be influenced by its immediate surroundings, and that interactions mediated by physical fields could not propagate faster than the speed of light. "Hidden variables" were properties of quantum particles that were not included in quantum theory but nevertheless affect the outcome of experiments. Bell deduced that if measurements were performed independently on the two separated particles of an entangled pair, then the assumption that the outcomes depended upon hidden

variables within each half implied a mathematical constraint on how the outcomes on the two measurements were correlated. This constraint was subsequently called a *Bell inequality*. Bell then showed that quantum physics predicts correlations that violate this inequality.

The first rudimentary experiment designed to test *Bell's theorem* was performed in 1972 by John Clauser and Stuart Freedman, and reported in Freedman, S. J. & Clauser, J. F. (1972). *Experimental Test of Local Hidden-Variable Theories*. "In the present work we measured the correlation in linear polarization of two *photons* emitted in an atomic cascade. The decaying atoms were viewed by two symmetrically placed optical systems, each consisting of two lenses, a wavelength filter, a rotatable and removable polarizer, and a single-photon detector. We made the following assumptions for any *local hidden-variable* theory: (1) The two *photons* propagate as separated localized particles. (2) A binary selection process occurs for each *photon* at each polarizer (transmission or no-transmission). This selection does not depend upon the orientation of the distant polarizer. In addition, we made the following assumption to allow a comparison of the *generalization of Bell's inequality* without experiment: (3) All *photons* incident on a detector have a probability of detection that is independent of whether or not the *photon* has passed through a polarizer. It has been shown by this generalization of *Bell's inequality* that the existence of local hidden variables imposes restrictions on this correlation in conflict with the predictions of quantum mechanics. Our data, in agreement with quantum mechanics, violated these restrictions to high statistical accuracy, *thus providing strong evidence against local hidden-variable theories*."

For this work in 1972, 50 years later, Clauser, jointly with Anton Zeilinger and Alain Aspect, was awarded the 2022 Nobel Prize in Physics, "for experiments with *entangled photons*, establishing the violation of Bell inequalities and pioneering quantum information science" … "One of the most remarkable traits of quantum mechanics is that it allows two or more particles to exist in what is called an *entangled state*. What happens to one of the particles in an *entangled* pair determines what happens to the other particle, even if they are far apart. In 1972, John Clauser conducted groundbreaking experiments using entangled light particles, photons. This and other experiments confirm that quantum mechanics is correct and pave the way for quantum computers, quantum networks and quantum encrypted communication."

8   The Standard Model

This is the fascinating story of the development of the *Standard Model* of particle physics between Dirac's prediction of the *positron* in 1928 and the introduction of the *six-quark model* in the mid 1970's, as described in the primary sources.

The *Standard Model of particle physics* is the theory describing three of the four known fundamental *forces (electromagnetic, weak and strong interactions* – excluding gravity) in the universe and *classifying all known elementary particles.*

[*Gravity*, also known as *gravitation* or a *gravitational interaction*, is a mutual attraction between all particles with mass.

Electromagnetism is an interaction that occurs between particles with electric charge via electromagnetic fields. The *electromagnetic force* is the dominant force in the interactions of atoms and molecules. Electromagnetism can be thought of as a combination of electrostatics and magnetism, which are distinct but closely intertwined phenomena. Electromagnetic forces occur between any two charged particles. Electric forces cause an attraction between particles with opposite charges and repulsion between particles with the same charge, while magnetism is an interaction that occurs between charged particles in relative motion. These two forces are described in terms of electromagnetic fields. Macroscopic charged objects are described in terms of Coulomb's law for electricity and Ampère's force law for magnetism; the Lorentz force describes microscopic charged particles.

The electromagnetic force is responsible for many of the chemical and physical phenomena observed in daily life. The electrostatic attraction between atomic nuclei and their electrons holds atoms together. Electric forces also allow different atoms to combine into molecules,

The *weak interaction*, *weak force* or the *weak nuclear force*, is the mechanism of interaction between subatomic particles that is responsible for processes such as the radioactive decay of atoms and the interactions between subatomic particles, particularly involving left-handed fermions and right-handed antifermions. The weak interaction participates in nuclear fission and nuclear fusion; this force plays a crucial role in nuclear fusion reactions, such as those that power the Sun. The weak interaction is mediated by the exchange of W and Z bosons, which are massive particles, leading to a very short effective range limited to subatomic distances of about 10^{-18} meters, less than the diameter of a proton.

The *strong interaction*, also called the *strong force* or *strong nuclear force*, confines quarks into protons, neutrons, and other hadron particles, and also binds neutrons and protons to create atomic nuclei, where it is called the nuclear force. Most of the mass of a proton or neutron is the result of the strong interaction energy; the individual quarks provide only about 1% of the mass of a proton. At the range of 10^{-15} m (1 femtometer, slightly more than the radius of a nucleon), the strong force is approximately 100 times as strong as electromagnetism, 10^6 times as strong as the weak interaction, and 10^{38} times as strong as gravitation.

In the context of atomic nuclei, the strong force binds protons and neutrons together to form a nucleus and is called the *nuclear force* (or *residual strong force*). Because the force is mediated by massive, short-lived mesons on this scale, the *residual strong interaction* obeys a distance-dependent behavior between nucleons that is quite different from when it is acting to bind quarks within hadrons. There are also differences in the binding energies of the nuclear force with regard to nuclear fusion versus nuclear fission. Nuclear fusion accounts for most energy production in the Sun and other stars. Nuclear fission allows for decay of radioactive elements and isotopes, although it is often mediated by the *weak interaction*. Artificially, the energy associated with the nuclear force is partially released in nuclear power and nuclear weapons, both in uranium or plutonium-based fission weapons and in fusion weapons like the hydrogen bomb.]

It was developed in stages throughout the latter half of the 20th century, through the work of many scientists worldwide, with the current formulation being finalized in the mid-1970s upon experimental confirmation of the existence of *quarks*. Since then, proof of the *top quark* (1995), the *tau neutrino* (2000), and the *Higgs boson* (2012) added further credence to the *Standard Model*. In addition, the *Standard Model* has predicted various properties of *weak neutral currents* and the W and Z *bosons* with great accuracy.

The Standard Model is a paradigm of a *quantum field theory*, exhibiting a wide range of phenomena, including *spontaneous symmetry breaking, anomalies,* and *non-perturbative behavior*. It is used as a basis for building more exotic models that incorporate hypothetical particles, extra dimensions, and elaborate *symmetries* (such as *supersymmetry*) to explain experimental results at variance with the *Standard Model*, such as the existence of dark matter and neutrino oscillations.

Although the *Standard Model* is believed to be theoretically self-consistent there are mathematical issues regarding *quantum field theories* still under debate. It has demonstrated some success in providing experimental predictions, but *it leaves some physical phenomena unexplained and so falls short of being a complete theory of fundamental interactions.* Although the physics of *special relativity* is included, *general relativity is not*, and it will fail at energies or distances where the *graviton* is expected to emerge. It does not fully explain *baryon asymmetry*, or account for the *universe's accelerating expansion as possibly described by dark energy.* The model does *not contain any viable dark matter particle* that possesses all of the required properties deduced from observational cosmology. It also does not incorporate *neutrino oscillations* and their non-zero masses.

This analysis of the failures of the *Standard Model* suggest that they stem from the attempt to base it on *relativistic quantum field theory* and make it *Lorentz covariant*, and the reliance on *renormalization* to remove infinities and bring theoretical values in line with experimental ones. (See Underwood, T. G. (2023). *Quantum Electrodynamics – annotated sources, Vol. II.*)

> [A physical quantity is said to be *Lorentz covariant* if it transforms under a given representation of the Lorentz group. According to the representation theory of the Lorentz group, these quantities are built out of scalars, four-vectors, four-tensors, and spinors. Theories that are *consistent with the principle of Special Relativity* must have the same form in all Lorentz frames, that is, they must be *covariant*. The *covariant* formulation of *classical electromagnetism* refers to ways of writing the laws of *classical electromagnetism* in a form that is manifestly invariant under *Lorentz transformations*.]

Part I describes the development of the *Standard Model* from the Bohr model of the atom in 1913, based on what were believed to be 3 stable particles, *electrons* in orbit around a *nucleus* comprised of *protons* and *neutrons*, to its emergence in 1973 as the *six-quark model*, comprising 52 elementary particles and anti-particles, revealed largely by tracks in cloud chambers from high energy collisions between particles, or at high altitudes with cosmic rays, of which only the electron is stable.

It starts with an overview of the Standard Model. This is followed by introductions to the four "fundamental interactions or forces" and the role of symmetry in fundamental physics, which describes the existence of symmetries in classical dynamics and in quantum mechanics, on which the Standard Model is based. This is followed by a summary of the timeline of the discovery of subatomic particles; a valuable resource which describes the large number *elementary and composite particles* and their *classifications*.

The *classifications* include *fermions* (elementary and composite particles with ½ integer spin (including quarks and leptons); *leptons* (elementary particles with ½ integer spin), including *muons* (charged leptons) and *neutrinos* (neutral leptons); *bosons* (elementary and composite particles with integer spin), including *vector bosons* (spin 1), of which *gauge bosons* are elementary particles which act as force carriers) including *gluons* (which carry the strong interaction); *hadrons* (composite particles made of two or more quarks held together by the strong interaction), including *baryons* (composite particles that contains an odd number of valence quarks and antiquarks) and *mesons* (composite particles that contain an equal number of quarks and antiquarks, usually one of each, bound together by the strong interaction); *hyperons* are baryons containing one or more strange quarks, but no charm, bottom, or top quark.

Quarks are elementary particles and a fundamental constituent of matter, which combine to form *hadrons*, the most stable of which are *protons* and *neutrons*, the components of atomic nuclei; they are the only elementary particles in the *Standard Model* to experience all four fundamental interactions, also known as fundamental forces (*electromagnetism, gravitation, strong interaction,* and *weak interaction*), as well as the only known particles whose *electric charges* are not integer multiples of the elementary *charge*. There are six types (*flavors*) of *quarks*: *up, down, charm, strange, top,* and *bottom.*

The *Standard Model* includes 26 *elementary* particles: 12 *fermions* (6 *quarks* and 6 *leptons*) and 14 *bosons* (13 *gauge bosons* and 1 *scalar boson,* the *Higgs boson*). The 6 *quarks* comprise *up, down, charm, strange, top* and *bottom*; the 6 *leptons* comprise the *electron, electron neutrino, muon, muon neutrino, tau* and *tau neutrino*; and the 13 *gauge bosons* comprise the *photon,* 2 *W bosons,* the *Z boson,* 8 types of *gluon,* and the hypothetical *graviton.* Each has its own anti-particle, making a total of 52 *elementary particles and anti-particles.*

In his 1928 article [Dirac, P. A. M. (February, 1928). *The Quantum Theory of the Electron*] and his 1933 Nobel Lecture [*Theory of electrons and positrons*], Dirac describes how by subjecting quantum mechanics to *relativistic requirements* he was able to deduce the existence and properties of the *antiparticle* of the *electron,* the *positron,* a particle with the same *mass* but opposite *electric charge*. He noted that *there was a complete and perfect symmetry between positive and negative electric charge.*

In 1932 and 1933, Heisenberg published a three-part paper on atomic nuclei which concluded by treating *protons* and *neutrons* on an equal footing by *considering protons and neutrons as different charge states of the same particle, which he referred to as the isotopic spin parameter.* [Heisenberg, W. (January, 1932). *Über den Bau der Atomkerne. I.* (About the construction of atomic nuclei. I.); (March, 1932). *Über den Bau der Atomkerne. II.* (About the construction of atomic nuclei. II.); (September, 1933). *Über den Bau der Atomkerne. III.* (About the construction of atomic nuclei. III.)]

In 1932, from a cloud chamber photograph of cosmic rays, the American physicist Carl David Anderson identified a track as having been made by a *positron.* [Anderson, C. D. (1933). *The Positive Electron.*]

Dirac [(April, 1934). *Discussion of the infinite distribution of electrons in the theory of the positron.*], addressed this with his *relativistic* 'hole' theory which implied an infinite number of negative-energy *electrons* (per unit volume) with energies extending continuously from $-mc^2$ to $-\infty$, so that when an electromagnetic field is present positive- and negative-energy states cannot be distinguished in a *relativistically* invariant way. He saw the need to set up assumptions for production of *electromagnetic field* by the electron distribution such that any finite change in distribution produces a change in the field in agreement with Maxwell's equations and such that the infinite field which would be required by Maxwell's equations from an infinite density of electrons is in some way cut out. Dirac assumed that each *electron* has its own individual wave function in space-time and moved in an *electromagnetic field* which was the same for all *electrons* part coming from external causes and part from the *electron distribution* itself. He then introduced a *relativistic density matrix* referring to two points in space and two times, and separated the density distribution into two parts where one, R_a contained the *singularities*, and the other, R_b described the *electric* and *current densities* physically present, so that any alteration in the distribution of *electrons* and *positrons* would correspond to an alteration in R_b, which was *relativistically invariant and gauge invariant*, and the *electric density* and *current density* corresponding to it satisfied the conservation law.

> [A *gauge theory* is a type of field theory in which the Lagrangian, and hence the dynamics of the system itself, do not change under *local* transformations according to certain smooth families of operations (Lie groups). Formally, the Lagrangian is invariant under these transformations.]

In this way, he *removed the infinities* and assumed that the *electric and current densities corresponding to R_b were those which were physically present, arising from the distribution of electrons and positrons*.

Heisenberg responded by presenting his thinking on Dirac's theory and further development of the theory in two papers [Heisenberg, W. (September, 1934). *Bemerkungen zur Diracschen Theorie des Positrons.* (Remarks on the Dirac theory of positron.); and Heisenberg, W., & Euler, H. (1936). *Folgerungen aus der Diracschen Theorie des Positrons.* (Consequences of Dirac's theory of the positron.)].

In these papers Heisenberg was the first to reinterpret the Dirac equation as a "classical" field equation for any point particle of *spin* $\hbar/2$, itself subject to quantization conditions involving anti-commutators. Thus, reinterpreting it as a *quantum field equation* accurately describing *electrons*, Heisenberg put matter on the same footing as *electromagnetism*: as being described by *relativistic quantum field equations* which allowed the possibility of particle creation and destruction.

In their 1936 paper, [Heisenberg, W., & Euler, H. (1936). *Folgerungen aus der Diracschen Theorie des Positrons.* (Consequences of Dirac's theory of the positron.)] Heisenberg and Euler noted that according to Dirac's *theory of the positron*, an *electromagnetic field* tends to create pairs of particles which leads to a change of Maxwell's equations in the vacuum. Their paper examined the consequence of the possibility of transforming *electromagnetic radiation* into *matter* in *quantum electrodynamics* on Maxwell equations. It is no longer possible to separate processes in the vacuum from those involving *matter* since *electromagnetic fields* can create *matter* if they are strong enough. Even if they are not strong enough to create *matter* they will, due to the virtual possibility of creating *matter*, *polarize the vacuum.*

In Fermi, E. (1934) *Versuch einer Theorie der β-Strahlen. I.* (Attempt at a theory of β rays. I.), Enrico Fermi proposed the first theory of the *weak interaction*, known as *Fermi's interaction*. He proposed a quantitative theory of *β decay* in which the existence of the *neutrino* was assumed, and the emission of *electrons* and *neutrinos* from a nucleus in the β case was treated with a method similar to that of the emission of a quantum of light from an excited atom in radiation theory. Formulas for the lifetime and for the shape of the emitted continuous β radiation spectrum are derived and compared with experience.

In Yukawa, H. (1935). *On the Interaction of Elementary Particles.*, Hideki Yukawa predicted the existence of *mesons* in his *theory of mesons* that postulated the particle as mediating the *nuclear force.* He noted that the interaction between *elementary* particles could be described by means of a *field of force*, just as the interaction between the *charged* particles is described by the *electromagnetic field.* He suggested that in quantum theory this field should be accompanied by a new sort of quantum, just as the electromagnetic field is accompanied by the *photon.* In this paper the possible natures of this field and the quantum accompanying it were discussed and their bearing on the nuclear structure considered.

In Wigner, E. (1937). *On the Consequences of the Symmetry of the Nuclear Hamiltonian on the Spectroscopy of Nuclei.*, Eugene Wigner investigated the structure of the *multiplets* of nuclear terms using as *first approximation* a Hamiltonian which does not involve the ordinary *spin* and corresponds to equal forces between all nuclear constituents, *protons* and *neutrons.*

[A *multiplet* is the *state space* for 'internal' degrees of freedom of a particle, that is, degrees of freedom associated to a particle itself, as opposed to 'external' degrees of freedom such as the particle's position in space. Examples of such degrees of freedom are the *spin state* of a particle in *quantum mechanics*, or the *color, isospin* and *hypercharge state* of particles in the *Standard Model*. Formally, this state space is described by a *vector space* which carries the *action* of a group of *continuous symmetries*.]

The *multiplets* turn out to have a rather complicated structure, instead of the S of atomic spectroscopy, one has three quantum numbers S, T, Y. The *second approximation* can either introduce *spin* forces (method 2), or else can discriminate between *protons* and *neutrons* (method 3). The *last approximation* discriminates between *protons* and *neutrons* as in method 2 and takes the *spin* forces into account as in method 3. The method 2 is worked out schematically and is shown to explain qualitatively the table of *stable nuclei* to about Mo.

Sakata and Inoue proposed their *two-meson theory* in 1942. [Sakata, S. & Inoue, T. (1942). *Chukanshi to Yukawa ryushi no Kankei ni tuite.* (in Japanese). (On the Correlations between Mesons and Yukawa Particles.)]. At the time, a charged particle discovered in the hard component cosmic rays was misidentified as the Yukawa's *meson* ($\pi^{\pm}$, nuclear force carrier particle). The misinterpretation led to puzzles in the discovered cosmic ray particle. Sakata and Inoue solved these puzzles by identifying the cosmic ray particle as a daughter charged *fermion* produced in the $\pi^{\pm}$ decay. A new neutral *fermion* was also introduced to allow $\pi^{\pm}$ decay into *fermions*. We now know that these charged and neutral *fermions* correspond to the second-generation *leptons* μ and ν_{μ} in the modern language. They then discussed the decay of the *Yukawa particle*, $\pi^{+} \rightarrow \mu^{+} + \nu^{\mu}$. Sakata and Inoue predicted correct *spin* assignment for the *muon*, and they also introduced the second *neutrino*. They treated it as a distinct particle from the *beta decay neutrino*, and anticipated correctly the three body decay of the *muon*. As a result of World War II, the English printing of Sakata-Inoue's *two-meson theory* paper was delayed until 1946, one year before the experimental discovery of $\pi \rightarrow \mu\nu$ decay.

In Yang, C. N. & Mills, R. (1954). *Conservation of Isotopic Spin and Isotopic Gauge Invariance.*, Chen-Ning Yang and Robert Mills extended the concept of *gauge theory* for *abelian* groups, e.g. *quantum electrodynamics*, to *nonabelian* groups to provide an explanation for *strong interactions*. The *Yang–Mills theory* is a *quantum field theory* for *nuclear binding*. It is a *gauge theory* based on a *special unitary group* SU(n), or more generally any compact Lie group.

> [The *special unitary group* of degree n, denoted SU(n), is the Lie group of n × n *unitary matrices* with *determinant* (real) 1.]

It seeks to describe the behavior of *elementary particles* using these non-abelian Lie groups and *is at the core of the unification of the electromagnetic force and weak forces* (i.e. U(1) × SU(2)) as well as *quantum chromodynamics*, the theory of the *strong force* (based on

SU(3)). Thus, it forms the basis of the understanding of the *Standard Model* of particle physics.

In 1956, [Sakata, S. (1956). *On a Composite Model for the New Particles.*] Sakata proposed his *Sakata Model* which explains the physics behind the *Nakano-Nishijima-Gell-Mann (NNG) rule* by postulating that the fundamental building blocks of all strongly interacting particles are the *proton*, the *neutron* and the *lambda baryon*. The positively charged *pion* is made out of a *proton* and an *anti-neutron*, in a manner similar to the *Fermi-Yang composite Yukawa meson model*, while the positively charged *kaon* is composed of a *proton* and an *anti-lambda*. Aside from the integer *charges*, the *proton*, *neutron*, and *lambda* have similar properties as the *up quark*, *down quark*, and *strange quark* respectively.

In 1956 [Lee, T. D. & Yang, C. N. (1956). *Question of Parity Conservation in Weak Interactions.*], Chen-Ning Yang and Tsung-Dao Lee formulated a theory that the *left-right symmetry law* is violated by the *weak interaction*. They noted that recent experimental data indicated closely identical *masses* and *lifetimes* of the θ^+ and the τ^+ mesons. On the other hand, analyses of the decay products of τ^+ strongly suggested on the grounds of *angular momentum* and *parity conservation* that the τ^+ and θ^+ were not the same particle. This posed a rather puzzling situation that has been extensively discussed. They suggested that one way out of the difficulty was to assume that *parity is not strictly conserved*, so that θ^+ and τ^+ are two different decay modes of the same particle, which necessarily has a single *mass* value and a single *lifetime*. They analyzed this possibility in this paper against the background of the existing experimental evidence of *parity conservation*, and concluded that existing experiments indicated *parity conservation* in *strong* and *electromagnetic interactions* to a high degree of accuracy, but that for the *weak interaction* actions (i.e., decay *interactions* for the *mesons* and *hyperons*, and various Fermi *interactions*) *parity conservation* was so far only an extrapolated hypothesis unsupported by experimental evidence. The question of *parity conservation* in β *decays* and in *hyperon* and *meson decays* was examined, and possible experiments were suggested which might test *parity conservation* in these *interactions*.

In Nambu, Y. (1960). *Axial Vector Current Conservation in Weak Interactions.*, Yoichiro Nambu noted that, in analogy to the conserved *vector current interaction* in the *beta decay* suggested by Feynman and Gell-Mann, some speculations have been made about a possibly conserved *axial vector current*. We would like to suggest that there may not be a strict *pseudovector current conservation*, but that we may have an approximate conservation which becomes rigorous in the limit $q^2 \gg m_\pi^2$, m_π being the *pion* mass and q^2 a *massless, pseudo scalar*, and *charged quantum* bridging the *nucleon* and *lepton currents*. We are tempted to extend this approximate conservation of the *axial vector current* (and naturally also the *vector current*) to the *strangeness-non conserving beta decays*.

116

In Sakurai, J. J. (1960). *Theory of strong interactions.*, Jun John Sakurai, a Japanese-American particle physicist, noted that all the *symmetry* models of *strong interactions* which had been proposed up to the present were devoid of deep physical foundations. He suggested that, instead of postulating artificial "higher" *symmetries* which must be broken anyway within the realm of *strong interactions*, we take the *existing exact* symmetries of *strong interactions* more seriously than before and exploit them to the utmost limit. A new theory of *strong interactions* was proposed on this basis.

In Goldstone, J. (1961). *Field theories with "Superconductor" solutions.*, Jeffrey Goldstone, a British theoretical physicist, examined the conditions for the existence of non-perturbative type "superconductor" solutions of field theories. A *non-covariant canonical transformation method* was used to find such solutions for a theory of a *fermion* interacting with a *pseudoscalar boson*. A *covariant renormalizable method* using Feynman integrals was then given. A *"superconductor" solution* was found whenever in the *normal perturbative-type solution* the *boson* mass squared was negative and the coupling constants satisfied certain inequalities. The *symmetry properties* of such solutions were examined with the aid of a simple model of self-interacting *boson* fields. The solutions had lower *symmetry* than the Lagrangian, and contain *mass zero bosons*.

Goldstone's theorem in *relativistic quantum field theory* states that if there is an exact *continuous symmetry* of the Hamiltonian or Lagrangian defining the system, and this is not a symmetry of the vacuum state (i.e. there is *broken symmetry*), then there must be at least one *spin-zero massless particle* called a *Goldstone boson*. In the *quantum theory of many-body systems Goldstone bosons* are collective excitations such as *spin waves*. An important exception to *Goldstone's theorem* is provided in *gauge theories* with the *Higgs mechanism*, whereby the *Goldstone bosons* gain *mass* and become *Higgs bosons*.

In Glashow, S. L. (1961). *Partial-symmetries of weak Interactions.*, Sheldon Glashow, Sheldon Glashow combined the *electromagnetic* and *weak interactions* and extended *electroweak unification models* due to Schwinger by including a short-range *neutral current*, the Z_0. The resulting *symmetry structure* that Glashow proposed, $SU(2) \times U(1)$, forms the basis of the accepted *theory of the electroweak interactions*. The *W and Z bosons* were predicted in detail by Sheldon Glashow, Mohammad Abdus Salam, and Steven Weinberg. For this discovery, Glashow along with Steven Weinberg and Abdus Salam, was awarded the 1979 Nobel Prize in Physics.

In Gell-Mann, M. (1961). *The Eightfold Way: A Theory of Strong Interaction Symmetry. (Report).*, Murray Gell-Mann introduced his formulation of a particle classification system for *hadrons* known as *the Eightfold Way* – or, in more technical terms, SU(3) *flavor symmetry*, streamlining its structure. A new model of the higher symmetry of elementary particles was introduced in which the eight known *baryons* were treated as a *super-*

117

multiplet, degenerate in the limit of *unitary symmetry* but split into isotopic *spin multiplets* by a *symmetry-breaking* term.

The *symmetry violation* was ascribed phenomenologically to the *mass* differences. The *baryons* corresponded to an eight-dimensional irreducible *representation* of the *unitary group*. The *pion* and *K meson* fit into a similar set of eight particles along with a predicted *pseudoscalar meson* χ^0 having I = 0. A ninth *vector meson* coupled to the *baryon* current could be accommodated naturally in the scheme. Gell-Mann predicted that the eight *baryons* should all have the same *spin* and *parity* and that *pseudoscalar* and *vector mesons* should form *octets* with possible additional *singlets*. The mathematics of the *unitary group* was described by considering three fictitious *leptons*, ν , e^-, and μ^-, which might throw light on the structure of *weak interactions*.

In Schwinger, J. (1962). *Gauge Invariance and Mass.*, Julian Schwinger argued that the *gauge invariance* of a *vector field* did not necessarily imply zero *mass* for an associated particle if the *current vector coupling* was sufficiently strong. He suggested that this situation might permit a deeper understanding of nucleonic *charge conservation* as a manifestation of a *gauge invariance*, without the obvious conflict with experience that a massless particle entailed.

In Goldstone, J., Salam, A. & Weinberg, S. (1962). *Broken Symmetries.*, some proofs were presented of Goldstone's Theorem, that "*in a manifestly Lorentz-invariant quantum field theory, if there is a continuous symmetry transformation under which the Lagrangian is invariant, then either the vacuum state is also invariant under the transformation, or there must exist spinless particles of zero mass*".

In Anderson, P. W. (1963). *Plasmons, Gauge Invariance, and Mass.*, Philip Anderson noted that Schwinger had pointed out that the *Yang-Mills vector boson* implied by associating a generalized *gauge transformation* with a conservation law (of *baryonic charge*, for instance) did not necessarily have zero *mass*, if a certain criterion on the vacuum fluctuations of the generalized *current* was satisfied. He showed that the theory of plasma oscillations was a simple *nonrelativistic* example exhibiting all of the features of Schwinger's idea. He also showed that Schwinger's criterion that the *vector field* $m \neq 0$ implied that the *matter* spectrum before including the *Yang-Mills interaction* contained $m = 0$, but that the example of *superconductivity* illustrated that the physical spectrum need not. Some comments on the relationship between these ideas and the zero-*mass* difficulty in theories with *broken symmetries* were given.

In Gell-Mann. M. (1964). *A Schematic Model of Baryons and Mesons.*, Murray Gell-Mann proposed that *baryons*, which include *protons* and *neutrons*, and *mesons* were composed of elementary particles, which Gell-Mann called "*quarks*". The theory came to be called the *quark model*. The bootstrap model for *strongly interacting particles* described in terms of the *broken eightfold way* was discussed to determine algebraic properties of the interactions with scattering amplitudes on the mass shell. A mathematical model based on *field theory* was described.

In 1964, James Cronin and Val Fitch, American particle and nuclear physicists, demonstrated, using *neutral K-mesons*, the violation of all three symmetry principles (1) that the laws of Nature are exactly alike for both *antimatter* and ordinary *matter*; (2) that the fundamental laws have exact *mirror symmetry*; and (3) that the fundamental laws have exact *time reflection symmetry* – symmetry under motion reversal. For this discovery, they received the 1980 Nobel Prize in Physics.

In Englert, F. & Brout, R. (1964). *Broken Symmetry and the Mass of Gauge Vector Mesons.*, François Englert and Robert Brout showed that *gauge vector fields*, abelian and non-abelian, could acquire *mass* if empty space were endowed with a particular type of structure that one encounters in material systems. Other physicists, Peter Higgs and Gerald Guralnik, C. R. Hagen and Tom Kibble had reached similar conclusions at about the same time. The *Brout–Englert–Higgs* (BEH) *mechanism* is believed to give rise to the *masses* of all the elementary particles in the *Standard Model*. This includes the *masses* of the W and Z *bosons*, and the *masses* of the *fermions*, i.e. the *quarks* and *leptons*.

In Higgs, P. W. (1964). *Broken Symmetries and the Masses of Gauge Bosons.*, Peter Higgs observed that in a recent note he had shown that the *Goldstone theorem*, that Lorentz-covariant field theories in which spontaneous breakdown of *symmetry* under an internal Lie group occurs contain *zero-mass* particles, *fails if and only if the conserved currents associated with the internal group are coupled to gauge fields*. The purpose of the present note was to report that, *as a consequence of this coupling, the spin-one quanta of some of the gauge fields acquire mass*; the longitudinal degrees of freedom of these particles (which would be absent if their *mass* were zero) go over into the Goldstone bosons *when the coupling tends to zero. The model was discussed mainly in classical terms*; nothing was proved about the quantized theory. Higgs commented that it should be understood, therefore, that *the conclusions which were presented concerning the masses of particles were conjectures based on the quantization of linearized classical field equations.*

In Salam, A. & Ward, J. C. (1964). *Electromagnetic and Weak Interactions.*, Abdus Salam and John Ward worked on the synthesis of the *weak* and *electromagnetic interaction*, obtaining a *gauge theory* based on the SU(2) × U(1) model.

In Weinberg, S. (1967). *Model of Leptons.*, Steven Weinberg incorporated the *Brout–Englert–Higgs (BEH) mechanism* into Glashow's *electroweak interaction*, giving it its modern form. Weinberg proposed his *model of unification of electromagnetism and nuclear weak forces* with the *masses* of the force-carriers of the *weak* part of the *interaction* being explained by *spontaneous symmetry breaking*, in which the *symmetry* between the *electromagnetic* and *weak interactions* was *spontaneously broken*, but in which the *Goldstone bosons* were avoided by introducing the *photon* and the *intermediate boson fields* as *gauge fields*.

In 't Hooft, G. (1971). *Renormalizable Lagrangians for Massive Yang-Mills Fields.*, Gerard 't Hooft showed how *renormalizable* models are constructed in which *local gauge invariance* was *broken spontaneously*. He noted that *Feynman rules* and *Ward identities* could be found by means of a path integral method, and they could be checked by algebra. In one of these models, which was studied in more detail, *local* SU(2) was broken in such a way that *local* U(1) remained as a *symmetry*. This resulted in a *renormalizable and unitary theory*, with *photons*, *charged massive vector particles*, and additional *neutral scalar particles*. It had three independent parameters. Another model had local SU(2) x U(1) as a *symmetry* and might serve as a *renormalizable* theory for ρ-*mesons* and *photons*. In such models, *electromagnetic mass-differences* were finite and could be calculated in *perturbation theory*.

In Kobayashi, M. & Maskawa, T. (1973). *CP-Violation in the Renormalizable Theory of Weak Interaction.*, Makoto Kobayashi and Toshihde Maskawa explained *broken symmetry* within the framework of the *Standard Model*, but *required that the Model be extended to three families of quarks to explain CP violation*, which ultimately led to the *six-quark model*.

Part II introduces an alternative, *Supersymmetry*; and **Part III** introduces, *String Theory*, neither of which fare any better.

9   New Physics   October 7, 2024

The exhaustive review of the primary sources of theoretical physics undertaken in my previous books† has revealed major problems which persist to the present day.

> † (April 2023) *Quantum Electrodynamics – annotated sources,* Volumes I and II; (June 2023) *Special Relativity*; (November 2023) *General Relativity*; (March 2024) *Gravity*; (May 2024) *Electricity & Magnetism*; (June 2024) *Quantum Entanglement*; (September 2024) *The Standard Model*.

These can be seen to be largely related to inconsistencies between *Einstein's theories of Special and General Relativity* and *quantum mechanics* and the consequent inability to quantize Einstein's *relativistic field equations*. Of particular concern is the fact that most of the so-called *elementary particles* in the *Standard Model* are largely derived from extremely high energy collisions between *protons*, have very short *half-lives*, between two one millionths and less than one million billion billionth of a second, and have *masses* derived almost entirely from *interaction energy*, making the *Standard Model* appear more like a theory of mass creation in high energy physics than a theory of *elementary particles*. As Dirac noted in his 1933 Nobel Lecture: "To get an interpretation of some modern experimental results one must suppose that particles can be created and annihilated. Thus, *if a particle is observed to come out from another particle, one can no longer be sure that the latter is composite. The former may have been created.*"

These problems include the following:

(1) Einstein's *theory of special relativity*. See Underwood, T. G. (June 2023). *Special Relativity*.

There is no evidence for Einstein's *theory of special relativity*, based directly or indirectly, on the observation of the speed of electromagnetic radiation in a vacuum emitted by an inertial body, or as observed by an inertial observer, moving in a straight line and not involving mirrors. However, by now it may be possible to achieve this in a laboratory experiment in a vacuum without mirrors, using electromagnetic radiation emitted by two sources of the same frequency, one stationary and the other moving at a constant velocity in a straight line; either directly, or by measuring the observed frequency of the radiation.

The Ehrenfest paradox, the *non-relativistic* Doppler red shift and blue shift for light, the known physics of the emission of electromagnetic radiation and of the electron, and the success of *non-relativistic* quantum electrodynamics in explaining the interaction of the electromagnetic field with electrically charged particles, comprise the strongest evidence against Einstein's *second postulate*, the *constancy of the speed of light*.

Quite apart from the enormity of the consequences of Einstein's two postulates taken together, including *length contraction*, *time dilation*, and the requirement to assume a *point electron* in the unsuccessful attempt to introduce special relativity into quantum electrodynamics, the evidence in support of Einstein's *second postulate* on the constancy of the speed of light is far outweighed by the evidence against it.

(2) Einstein's *theory of general relativity*. See Underwood, T. G. (November 2023). *General Relativity.*

Einstein's *theory of general relativity* attempted to extend his *theory of special relativity* beyond space and time, to include *matter* and *gravitational fields*. Whilst this allowed Einstein to construct a *relativistic theory* of the effect of a *gravitational field* on *matter*, it also resulted in him rejecting his *postulate on the constancy of light* in the presence of a gravitational field.

General relativity is claimed to generalize *special relativity* and refine Newton's *law of universal gravitation*, providing a unified description of *gravity* as a geometric property of *space and time* or four-dimensional *spacetime*. In particular, the *curvature of spacetime* is directly related to the *energy* and *momentum* of whatever *matter* and *radiation* are present. The relation is specified by the *Einstein field equations*, a system of second-order partial differential equations.

In order to make calculations with his theory, Einstein had to import *Newton's law of gravitation*, which itself is an empirical law with no fundamental foundation. Consequently, the only evidence that Einstein could provide for his *theory of general relativity* was effectively Newtonian.

Einstein, A. (1917). *Kosmologische Betrachtungen zur allgemeinen Relativitätstheorie.* (Cosmological Considerations in the General Theory of Relativity.) describes Einstein's struggles with supplementing the *relativistic differential equations* by *limiting conditions* at *spatial infinity* in order to regard the universe as being of infinite spatial extent. As he noted, "we admittedly had to introduce an extension of the field equations of gravitation which is not justified by our actual knowledge of gravitation".

(3) Gravity. See Underwood, T. G. (March 2024). *Gravity.*

Reconciliation of *general relativity* with the laws of *quantum physics* remains a problem *as there is a lack of a self-consistent theory of quantum gravity*. It is not yet known how gravity can be unified with the three non-gravitational forces: strong, weak and electromagnetic.

(4) Elementary and composite particles with the same electric charge attract each other, and elementary and composite particles with opposite electric charge are repulsed, through the electromagnetic interaction or electromagnetic force, according to Coulomb's law. See Underwood, T. G. (May 2024). *Electricity & Magnetism*.

The Standard Model adds nothing to the classical *non-relativistic* theory.

(5) Quantum field theory (relativistic quantum electrodynamics) and renormalization. See Underwood, T. G. (April 2023). *Quantum Electrodynamics – annotated sources. Volume II.*

The lack of convergence in current formulations of *relativistic quantum electrodynamics* for the *electron*, or *quantum field theory*, due to the interaction of the electromagnetic and matter fields with their own vacuum fluctuations raised the question of whether the still unresolved *divergencies* arising largely, if not entirely, from the assumption of a *point electron*, could be isolated in unobservable *renormalization* factors.

> [Underwood, T. G. (April 2023). *Quantum Electrodynamics – annotated sources. Volume II.* Preface, pp. 34-35: "Schwinger, in the Preface of his 1958 book [*Selected Papers on Quantum electrodynamics*], "questioned whether *renormalization* simply corrected a mathematical error that causes the divergencies, or whether *there is a serious flaw in the structure of field theory*". Feynman, in his 1965 Nobel prize speech, described *renormalization* as "simply a way to sweep the difficulties of the divergences of electrodynamics under the rug". Dirac's final judgment on *quantum field theory*, in his last paper published in 1987 [*The inadequacies of quantum field theory.*], was that "These rules of *renormalization* give surprisingly, excessively good agreement with experiments. Most physicists say that these working rules are, therefore, correct. I feel that is not an adequate reason. Just because the results happen to be in agreement with observation does not prove that one's theory is correct."]

Despite the claims to the contrary in modern textbooks, there have been no significant developments in the quantum electrodynamics or quantum field theory since 1965 to resolve the underlying occurrence of divergencies.

The *standard model* was established in the 1970s. It was triggered by the development of studies of *gauge theories*. In particular, it was proved that a generalized *gauge theory* is *renormalizable*.

123

[A *gauge theory* is a type of field theory in which the Lagrangian, and hence the dynamics of the system itself, do not change under *local* transformations according to certain smooth families of operations (Lie groups). Formally, the Lagrangian is invariant under these transformations.]

This opened the possibility that all the *interactions* of an *elementary particle* could be described by the *quantum field theory* without the difficulty of *divergence*. Before this time, such description was possible only for *electro-magnetic interaction*.

(6) Elementary particles. See Underwood, T. G. (September 2024). *The Standard Model.*

The *Standard Model* comprises a total of 52 *elementary particles* and their *anti-particles* of which only the *electron* and the *photon* are stable. Most of these have been revealed largely by tracks in cloud chambers from high energy collisions between particles in *proton-antiproton* and *hadron* colliders, or at high altitudes with cosmic rays. The *half-lives* of the other *elementary particles* vary between 2.2×10^{-6} s for the *muon*, to 3×10^{-25} s for the W^+, W^- and Z^0 bosons, 1.6×10^{-25} s for the Higgs boson, and 5×10^{-25} s for the top quark, particles with a *very high mass*.

(7) Masses. See Underwood, T. G. (September 2024). *The Standard Model.*

Most of the *mass* of a *proton* or *neutron* is the result of the *strong interaction* energy; the individual *quarks* provide only about 1% of the *mass* of a *proton*. *Protons* are composed of two *up quarks* (*mass* of each equal to 0.002 of *proton mass*), and one *down quark* (*mass* equal to 0.005 of *proton mass*). *Neutrons* are composed of two *down quarks*, and one *up quark*.

The large *masses* of the W^+, W^-, Z^0 and *Higgs bosons* and the *top quark* (respectively 85.7, 85.7, 97.2, 133.3 and 184.9 times the *mass* of the *proton*) relative to the *masses* of the *proton* and *neutron* raise questions regarding whether they are really *elementary particles*, in particular in view of how they were created.

(8) Symmetries. See Underwood, T. G. (September 2024). *The Standard Model.*

The Standard Model is a *gauge quantum field theory* containing the *internal* (*local*) *symmetries* of the *unitary product group* SU(3) × SU(2) × U(1). Roughly, the three factors of the gauge symmetry give rise to the three fundamental interactions, the *strong, weak* and *electromagnetic interactions*.

The three local symmetries addressed by the Standard Model are:
C-symmetry (charge symmetry), a universe where every particle is replaced with its antiparticle; *P-symmetry* (parity symmetry), a universe where everything is mirrored along

the three physical axes. This excludes weak interactions; *T-symmetry* (time reversal symmetry), a universe where the direction of time is reversed.

CP violation, the violation of the combination of *C-* and *P-symmetry*, is necessary for the presence of significant amounts of *baryonic matter* in the universe.

(9) Spin. See Underwood, T. G. (September 2024). *Quantum Entanglement.*

The *spin* of an *elementary particle* is a quantum state and consequently a *non-relativistic* concept. We could try to determine the behavior of *spin* under general Lorentz transformations, but we would immediately discover a major obstacle. Unlike SO(3), the group of Lorentz transformations SO(3,1) is *non-compact* and therefore does not have any faithful, unitary, finite-dimensional representations.

(10) Exchange interaction. See Underwood, T. G. (September 2024). *The Standard Model.*

In the *Standard Model of particle physics* four *fundamental interactions* or *forces* are assumed: *gravity*, and the *electromagnetic, weak and strong interactions*, of which the latter three are incorporated in the model.

[*Gravity*, also known as *gravitation* or a *gravitational interaction*, is a mutual attraction between all particles with mass.

Electromagnetism is an interaction that occurs between particles with electric charge via electromagnetic fields. The *electromagnetic force* is the dominant force in the interactions of atoms and molecules.

The *weak interaction*, *weak force* or the *weak nuclear force*, is the mechanism of interaction between subatomic particles that is responsible for processes such as the radioactive decay of atoms and the interactions between subatomic particles. The weak interaction participates in nuclear fission and nuclear fusion; this force plays a crucial role in nuclear fusion reactions, such as those that power the Sun. The weak interaction is mediated by the exchange of W and Z bosons, which are massive particles, leading to a very short effective range limited to subatomic distances of about 10^{-18} meters, less than the diameter of a proton.

The *strong interaction*, also called the *strong force* or *strong nuclear force*, confines *quarks* into protons, neutrons, and other hadron particles, and also binds neutrons and protons to create atomic nuclei, where it is called the *nuclear force*. Most of the mass of a proton or neutron is the result of the strong interaction energy; the individual quarks provide only about 1% of the mass of a proton. In the context of atomic nuclei, the strong force binds protons and neutrons together to form a nucleus and is called the *nuclear force* (or *residual strong force*). Because the force binding protons and neutrons together is mediated by massive, short-lived *mesons*, the *residual strong interaction* obeys a distance-dependent behavior between nucleons that is quite different from when it is acting to bind quarks within hadrons.

At the range of 10^{-15} m (1 femtometer, slightly more than the radius of a nucleon), the strong force is approximately 100 times as strong as electromagnetism, 10^6 times as strong as the weak interaction, and 10^{38} times as strong as gravitation.]

According to the *quark formulation* in the *Standard Model*, a *weak interaction* occurs when two particles (typically, but not necessarily, half-integer spin fermions) exchange integer-spin, force-carrying *bosons*. In the *weak interaction, fermions* can *exchange* three types of force carriers, namely W+, W−, and Z *bosons*. The *weak interaction* is the only fundamental *interaction* that breaks *parity symmetry*, and similarly, but far more rarely, the only *interaction* to break *charge–parity symmetry*. The *weak interaction* is considered unique in that it allows *quarks* to *swap* their flavor for another. The *swapping* of those properties is mediated by the force carrier *bosons*.

Composite particles, such as *protons* and *neutrons*, can exist as different *quantum states*, referred to as *isospin states*, which create an attractive force, the *strong interaction*, through *exchange interaction* between two *quantum isospin states*.

[The name of the concept *isospin* contains the term *spin* because its quantum mechanical description is mathematically similar to that of *angular momentum* (in particular, in the way it *couples*; for example, a *proton–neutron pair* can be *coupled* either in a *state* of *total isospin* 1 or in one of 0, but unlike angular momentum, it is a dimensionless quantity and is not actually any type of spin.]

Before the concept of *quarks* was introduced, particles that are affected equally by the *strong force* but had different *electric charges* (e.g. *protons* and *neutrons*) were considered different states of the same particle, but having *isospin* values related to the number of *charge states*. A close examination of *isospin symmetry* ultimately led directly to the discovery and understanding of *quarks* and to the development of *Yang–Mills theory*.

(11) Quarks. See Underwood, T. G. (September 2024). *The Standard Model.*

Quarks, which make up composite particles like *neutrons* and *protons*, come in six "*flavors*" – *up*, *down*, *charm*, *strange*, *top* and *bottom* – which give those composite particles their properties.

The *top quark had a mass much larger than expected*, almost as large as that of a gold atom. It has a mass of 172.76 ± 0.3 GeV/c2, (185 times the mass of a proton), which is close to the rhenium atom mass. Because the *top quark* is so massive, *its properties allowed indirect determination of the mass of the Higgs boson*. As such, the *top quark*'s properties are extensively studied as a means to discriminate between competing theories of new physics beyond the *Standard Model. The top quark is the only quark that has been directly observed* due to its decay time being shorter than the hadronization time.

The model was discussed mainly in classical terms; nothing was proved about the quantized theory. Higgs noted that it should be understood, therefore, that *the conclusions which were presented concerning the masses of particles were conjectures based on the quantization of linearized classical field equations.*

(12) Antimatter. See Underwood, T. G. (September 2024). *The Standard Model.*

Particles with the same *mass* but opposite *electric charge* to an existing *particle* are described as *anti-particles*. In the Standard Model, they are a different form of *matter* known as *antimatter*. The *electric charge* of the *positron* (the *anti-electron*) is − e (i.e. − 1 *electron charge*, positive). *Quarks* have fractional *electric charges.*

According to this theory, there are compelling theoretical reasons to believe that, aside from the fact that *antiparticles* have different signs on all charges (such as *electric* and *baryon charges*), *matter* and *antimatter* have exactly the same properties. This means a *particle* and its corresponding *antiparticle* must have identical *masses* and *decay lifetimes*. It is claimed that the electron's antiparticle, the *positron*, is *stable*, but in condensed matter it typically remains only a short time (10^{-10} sec) before annihilating with an *electron*. Similarly, the *antiproton* is claimed to be *stable* but is short-lived due to collisions with *protons*.

The discovery of *CP violation* implies that there is an essential difference between *particles* and *anti-particles. Matter dominance* of the universe seems to require new sources of *CP violation* because it appears that *CP violation* of the *six-quark model* is too small to explain *matter* dominance.

> [CP violation is a violation of CP-symmetry (or charge conjugation parity symmetry): the combination of C-symmetry (*charge conjugation symmetry*) and P-symmetry (*parity symmetry*). CP-symmetry states that the laws of physics should be the same if a particle is interchanged with its antiparticle (C-symmetry) while its spatial coordinates are inverted ("mirror" or P-symmetry). CP violation is only observed in the weak interaction.]

According to the Standard Model, the asymmetry of *matter* and *antimatter* (*baryon asymmetry*) in the visible universe is one of the great unsolved problems in physics.

(13) Other shortcomings.

The *Standard Model* leaves some physical phenomena unexplained and so falls short of being a complete theory of fundamental interactions. Although the physics of special relativity is included, *general relativity* is not, and it will fail at energies or distances where the *graviton* is expected to emerge.

It does not account for the universe's accelerating expansion as possibly described by *dark energy*.

The model does not contain any viable *dark matter* particle that possesses all of the required properties deduced from observational cosmology.

It also does not incorporate *neutrino oscillations* and their *non-zero masses*.

(14) Supersymmetry and String theory. See Underwood, T. G. (September 2024). *The Standard Model.*

Supersymmetry could help explain certain phenomena, such as the nature of *dark matter* and the *hierarchy problem* in particle physics. *There is no experimental evidence that either supersymmetry or misaligned supersymmetry holds in our universe*, and *many physicists have moved on from supersymmetry and string theory entirely due to the non-detection of supersymmetry at the Large Hadron Collider* (LHC).

Because *string theory* potentially provides a unified description of *gravity* and *particle physics*, it is a candidate for a theory of everything, a self-contained mathematical model that describes all fundamental forces and forms of matter. Despite much work on these problems, *it is not known to what extent string theory describes the real world* or how much freedom the theory allows in the choice of its details.

Foundational assumptions

This book attempts to highlight these problems by presenting a formulation of the foundations of physics, which I refer to as "*New Physics*", in which they are largely avoided by replacing *Einstein's theory of Special Relativity*, in which the speed of light is constant for all observers *regardless of the motion of light source or observer*, with *Ritz's emission theory*, in which the speed of light is constant *with respect to the emitter*.

Without professing to be a complete theory this is intended to clarify these issues and ideally encourage an experiment to demonstrate directly whether the speed of light travelling in a straight line is in fact independent of the speed of the emitter.

Part I examines the foundational assumptions of the Standard Model and **Part II** sets out the corresponding foundational assumptions of *New Physics*. Both include the corresponding annotated primary source documents. In **Part III**, they are brought together under each heading to highlight the differences.

The foundational assumptions of the *Standard Model* are

1) The universe is composed of 52 *elementary particles* and *antiparticles*, of which only the electron and photon are stable. *Antiparticles*, which have *differences in quantum numbers* in additional to electric charge, form *antimatter*, a different form of matter. There is an unexplained asymmetry of *matter* and *antimatter* in the visible universe;

2) The speed of light in a vacuum is constant for all observers, regardless of the motion of light source or observer. This results in *length contraction* and *time dilation* for a moving observer;

3) All elementary particles, including electromagnetic waves (photons), are quantized, but there is a *lack of convergence* in current formulations of quantum electrodynamics due to the interaction of the electromagnetic and matter fields with their own vacuum fluctuations. The question is whether all divergencies can be isolated in unobservable *renormalization* factors;

4) *All elementary* particles have *mass*, apart from the photon and gluons, but the large *masses* of the W^+, W^-, Z^0 and *Higgs bosons* and the *top quark* (respectively 85.7, 85.7, 97.2, 133.3 and 184.9 times the *mass* of the *proton*) raise questions regarding whether they are really *elementary particles*, in particular in view of how they were created by collisions of high energy *protons* in *proton* and *hadron* colliders.

5) *Elementary* and *composite* particles can have *electric charge* or be *neutral*. *Up*, *charm*, and *top quarks* are assumed to have *electric charges* equal to 2/3 of the charge of the *electron* and *proton*; and *down, strange,* and *bottom quarks* to have *electric charges* equal to – 1/3 of the charge of the *electron* and *proton*, though this cannot be observed;

6) *Elementary* and *composite* particles have a *quantum state* called *spin*;

7) *Elementary* and *composite* particles with *mass* attract each other through the *gravitational interaction or gravitational force*, according to Einstein's theory of General Relativity;

8) *Elementary* and *composite* particles with the same *electric charge* attract each other, and *elementary* and *composite* particles with opposite *electric charge* are

repulsed, through the *electromagnetic interaction* or *electromagnetic force*, according to Coulomb's law;

9) The *spin* of *elementary* and *composite* particles creates an attractive force between two particles - the *weak interaction* or *weak force* – through *exchange interaction* when *quarks* exchange integer-spin, force carrying *bosons*. This is also ascribed to *weak isospin* and *weak hypercharge*;

10) *Composite* particles, such as *protons* and *neutrons*, exist as *quantum states* with different *baryon numbers*, referred to as *iso-spin states*, which creates an attractive force between two particles – the *strong interaction* or *strong force* – through *exchange interaction* between two *quantum iso-spin states*.

The foundational assumptions of *New Physics* are

1) The universe is composed of 14 *elementary particles*, of which the *electron*, *photon*, *proton*, and *neutron* in the nucleus, are stable. *Elementary particles* are confined to those that have the possibility of being *observed*. No *quarks*, *gluons*, nor possibly W^+, W^-, Z^0, or *Higgs bosons*. Elimination of notion of *antimatter* and problem of asymmetry of *matter* and *antimatter* in the visible universe. *Antiparticles* are simply the less stable *particles* of similar *mass* but of opposite *electric charge*.

2) The speed of light in a vacuum is constant relative to the emitter: replaces *Einstein's theory of Special Relativity* with *Walter Ritz's emission theory*; avoids *length contraction* and *time dilation* for a moving observer.

3) All *elementary* particles, including electromagnetic waves (photons), are quantized. Only relationships among observable quantities occur. Avoids requirement to assume a *point electron* and address through a process of *renormalization* the still unresolved *divergencies*;

4) All *elementary* particles have *mass* apart from the *photon*. Possible removal of W^+, W^-, Z^0 and *Higgs bosons* and the *top quark* would avoid problem of overweight particles;

5) *Elementary* and *composite* particles can have *electric charge* or be neutral. Removal of *quarks* avoids the problem of elementary particles with unobserved fractional *electric charges*;

6) *Elementary* and *composite* particles have a *quantum state* called spin. *Spin* obeys the mathematical laws of angular momentum quantization;

7) *Elementary* and *composite* particles with *mass* attract each other through the *gravitational interaction* or *gravitational force* resulting from *quantum entanglement between matter*. Introduction of a *quantum theory of gravity*;

8) *Elementary* and *composite* particles with the same *electric charge* attract each other, and *elementary* and *composite* particles with opposite *electric charge* are

repulsed, through the *electromagnetic interaction* or *electromagnetic force*, according to Coulomb's law;

9) The *spin* of *elementary* and *composite* particles creates an attractive force between two particles - the *weak interaction* or *weak force* – resulting from *quantum entanglement* between *spin states*;

10) *Elementary* particles, such as *protons* and *neutrons*, and *composite* particles, can exist as *quantum states* referred to as *iso-spin states*, which creates an attractive force between two particles – the *strong interaction* or *strong force* – resulting from *quantum entanglement* between *isospin states*.

10   Cosmological Redshift of Light November 30, 2024

This book addresses the cause of the *cosmological redshift of light*. It is currently assumed to be due to the *Doppler effect* on light resulting from the *expansion of the universe* following the *Big Bang*. However, there is an alternative less radical theory. Fritz Zwicky's *"tired-light" theory*, attributes the *linear redshift with distance from the observer* to the *loss of energy by photons*, and consequent increase in *wavelength*, resulting from *interactions between the photons and intervening electrons or matter* whilst travelling through *intergalactic* space.

It is possible that both phenomena contribute to the observed redshift but if the *loss of energy by photons* is sufficient, this would suggest that *there was no Big Bang, the universe is not just 13.8 billion years old*, but *is indefinitely old, and in a steady state, not expanding*. In the absence of discussion by recognized institutions, the analysis below, that has been pieced together from disparate, largely non-mainstream, sources, suggests that is the case.

The notion of an *expanding universe* was first scientifically originated by physicist Alexander Friedmann in 1922, a Russian cosmologist and mathematician, who derived what became known as the Friedmann equations from the field equations of *Einstein's theory of general relativity*, showing that the universe might be expanding in contrast to the static universe model advocated by Albert Einstein at that time. [Friedman, A.A. (1922). *Über die Krümmung des Raumes.* English translation in Friedman, A.A. (1999). *On the curvature of space. General Relativity and Gravitation.*]

Independently deriving Friedmann's *relativistic* equations, in 1927, the year he became professor of astrophysics at the Catholic University of Leuven (Louvain), Georges Lemaître published in Belgium a virtually unnoticed paper that provided a compelling solution to the equations of General Relativity for the case of an expanding universe and proposed that the presumed recession of the nebulae was due to the *expansion of the universe*. He inferred the relation between the *redshift* and *distance* that Hubble would later observe.

In 1931, Lemaître went further and suggested that the evident *expansion of the universe*, if projected back in time, meant that the further in the past the smaller the universe was, until at some finite time in the past all the mass of the universe was concentrated into a single point, a "primeval atom" where and when the fabric of time and space came into existence, in what was later referred to as the *"Big Bang singularity"*.

In the 1920s and 1930s, almost every major cosmologist preferred an eternal *steady-state universe*. During the 1930s, other ideas were proposed as non-standard cosmologies to explain Hubble's observations, including Fritz Zwicky's *"tired-light" hypothesis*.

Subsequently, various cosmological models of the *Big Bang* attempted to explain the evolution of the observable universe from the earliest known periods through its subsequent large-scale form. They offered an explanation for a broad range of observed phenomena, including the abundance of light elements, the *cosmic microwave background* (CMB) radiation, and large-scale structure. The models depended on two major assumptions: the *universality of physical laws* (one of the underlying principles of the *theory of relativity*) and the *cosmological principle*. These models were compatible with the Hubble–Lemaître law—the observation that *the farther away a galaxy is, the faster it appeared to be moving away from Earth*. Under this theory, detailed measurements of the *redshift* placed the *Big Bang singularity* at an estimated 13.8 billion years ago, which was considered the *age of the universe*.

In 1929, Edwin Hubble published his famous paper [Hubble, E. (1929). *A relation between distance and radial velocity among extra-galactic nebulae.*] in which he used the strong direct relationship between a classical Cepheid variable's *luminosity* and *pulsation period* for scaling galactic and extragalactic *distances* and confirmed that the *redshift* of a galaxy, expressed as its *radial velocity*, increases with its *distance* from Earth, a behavior that became known as Hubble's law.

In response to Hubble's announcement of this somewhat linear relationship, Fritz Zwicky immediately pointed out that the correlation between the calculated *distances* of galaxies and their *redshifts* had a discrepancy too large to fit in the *distance*'s error margins. [Zwicky, F. (1929). *On the Red Shift of Spectral Lines through Interstellar Space.*] Zwicky *proposed his* "tired-light" theory that *the reddening effect was not due to motions of the galaxy, but to an unknown phenomenon that caused photons to lose energy as they traveled through space.* He considered the most likely candidate process to be a drag effect in which *photons* transfer *momentum* to surrounding masses through *gravitational interactions*. [However, *photons* have no *mass*.] He rejected explanations involving the *expansion of space*, and incorrectly stated that *Compton scattering* suffers from the blurring of images.

> [The MIT OpenCourseWare article, ***Interactions of Photons with Matter.***, describes three *energy loss mechanisms*, the *photoelectric effect; Compton scattering;* and *pair production*. It notes that because photons are electrically neutral, they *do not steadily lose energy* via coulombic interactions with atomic *electrons*, as do charged particles. Photons travel some considerable distance before undergoing a more *"catastrophic" interaction* leading to *partial or total transfer*

of the *photon* energy to *electron* energy. These *electrons* will ultimately deposit their energy in the medium. Photons are far *more penetrating* than *charged particles* of similar *energy*.

The *photoelectric* process is the *predominant mode of photon interaction* at *relatively low photon energies and high atomic number*. In the *photoelectric absorption process*, a *photon* undergoes an interaction with an *absorber atom* in which the *photon* completely disappears. In its place, an energetic *photoelectron* is ejected from one of the bound shells of the *atom*. For *gamma rays* of sufficient energy, the most probable origin of the *photoelectron* is the most tightly bound or K shell of the atom. For *gamma-ray* energies of more than a few hundred keV, the *photoelectron* carries off the majority of the original *photon energy*. The *photoelectric interaction* is most likely to occur if the energy of the incident *photon* is *just greater than the binding energy* of the *electron* with which it interacts.

Compton scattering, named after Arthur Compton an American physicist who won the Nobel Prize in Physics in 1927 for his 1923 discovery of the Compton effect, takes place between the incident *gamma-ray photon* and an *electron* in the absorbing material. In Compton scattering, the incoming *gamma-ray photon* is *deflected* through an angle θ with respect to its original direction. The *photon* transfers a portion of its *energy* to the *electron* (assumed to be initially at rest), which is then known as a *recoil electron*, or a *Compton electron*. All angles of scattering are possible. The *energy* transferred to the *electron* can vary from zero to a large fraction of the *gamma-ray energy*. The *Compton scattering* probability is almost *independent of atomic number*; decreases as the *photon energy* increases; is *directly proportional* to the number of *electrons* per gram, which only varies by 20% from the lightest to the heaviest elements (except for hydrogen).

If a *photon* enters matter with an energy *in excess of 1.022 MeV*, it may interact by a process called *pair production*. The *photon*, passing near the *nucleus* of an atom, is subjected to strong field effects from the *nucleus* and may disappear as a *photon* and reappear as a positive and negative *electron pair*. The two *electrons* produced, e⁻ and e⁺, are not scattered *orbital electrons*, but are created, *de novo*, in the *energy/mass conversion* of the disappearing *photon*. The *kinetic energy* of the *electrons* produced will be the difference between the *energy* of the incoming *photon* and the energy equivalent of two *electron* masses (2 x 0.511, or 1.022 MeV). *Pair production probability*, increases with *increasing photon energy*; and increases approximately as the square of the *atomic number*.

The *bulk behavior of photons in an absorber* is given by the *probability of an interaction per unit distance traveled*, $N = N_0 e^{-\mu x}$, where the linear *attenuation coefficient* μ, has the dimensions of inverse length (eg. cm^{-1}) and depends on *photon energy* and on the *material being traversed*; and the *mass attenuation coefficient*, μ/ρ, is obtained by dividing μ by the *density* ρ of the material, usually expressed in cm^{2}g^{-1}.]

Compton's paper [Compton, A. H. (1923). *A Quantum Theory of the Scattering of X-rays by Light Elements.*] confirmed that the scattering of *electrons* by *X-rays* or *γ-rays* results in a *redshift* due to the transfer of *energy* from the *photons* to the scattered *electron*. For *electromagnetic radiation* from remote *galaxies* observed from the Earth travelling through *intergalactic* space, known to contain *electrons* and other matter, this results in a *linear increase in the redshift with distance travelled by the photons*. This contributes to the *cosmological redshift of light*, which is currently assumed to be due to the *Doppler effect* on light resulting from the *expansion of the universe* following the *Big Bang*. It is possible that both phenomena contribute to the observed *redshift* but if the *loss of energy by photons* is sufficient, this would suggest that *there was no Big Bang, the universe is not just 13.8 billion years old*, but *is indefinitely old, and in a steady state, not expanding.*

Compton's paper noted that classical electrodynamics predicts that the *energy* scattered by an *electron* traversed by an *X-ray* beam is independent of the *wave-length* of the incident rays. It also predicts that when the *X-rays* traverse a thin layer of *matter* the *intensity* of the scattered radiation on the two sides of the layer should be the same. But experiments on the scattering of *X-rays* by light elements showed that these predictions were correct when *X-rays* of moderate hardness were employed, but when very hard *X-rays* or *γ-rays* were employed, the scattered energy was less than Thomson's theoretical value and was strongly concentrated on the emergent side of the scattering plate. Compton applied Einstein's hypothesis to the scattering of *X-ray* and *γ-ray photons* by *electrons* and derived the mathematical relationship between the shift in *wavelength* and the scattering angle of the *X-rays by assuming that each scattered X-ray photon interacts with only one electron.* This agreed with experimental results for the scattering of *X-ray* and *γ-ray photons* by *electrons*, subsequently known as *Compton scattering*, providing important evidence for *quantum theory*. [The introduction of *special relativity* was irrelevant to the comparison of the theory with experimental results.]

Burrows. A. S. (2015). *Baade and Zwicky: "Super-novae," neutron stars, and cosmic rays*, describes how, in 1934, two astronomers in two of the most prescient papers in the astronomical literature [Baade, W., & Zwicky, F. (1934*). On super-novae*. and Baade, W., & Zwicky, F. (1934). *Cosmic rays from super-novae.*] coined the term "*supernova*", hypothesized the existence of *neutron stars*, and knit them together with the origin of *cosmic-rays* to inaugurate one of the most surprising syntheses in the annals of science.

In Baade, W. (1938). *The Absolute Photographic Magnitude of Supernovae.*, Walter Baade provided a compilation of the *photometric data* for the 18 *supernovae* known at the end of 1937. Former estimates were replaced by *photometric magnitudes* after a redetermination of the magnitudes of comparison stars on the international system. The mean absolute *photographic magnitude* of the *supernovae*, derived from this material, was $\mathbf{M^-_{max}} = -14.3 \pm 0.42$ (m. e.) with a dispersion $M_{max} \approx 1.1$ mag. This result, together with the spectroscopic evidence, fully confirmed the view that two classes of *novae*, *common novae* and *supernovae*, exist. Attention was drawn to the curious fact that 72 per cent of the known *supernovae* appeared in late-type spirals. *B Cassiopeiae* and the *Crab nebula*, which may have been galactic *supernovae*, were discussed.

Tonry, J. & Schneider, D. P. (1988). *A New Technique for Measuring Extragalactic Distances.*, describes a relatively direct technique of determining *extragalactic distances*. The method relies on measuring the *luminosity fluctuations* that arise from the *counting statistics* of the *stars* contributing the *flux* in each pixel of a high-signal-to-noise CCD (*Charge Coupled Device*) image of a *galaxy*. *The amplitude of these fluctuations is inversely proportional to the distance of the galaxy.* This approach bypasses most of the successive stages of calibration required in the traditional *extragalactic distance ladder*; the only serious drawback to this method is that it requires an accurate knowledge of the bright end ($M_v < 3$) of the *luminosity function*. Potentially, this method can produce accurate distances of *elliptical galaxies* and *spiral bulges* at distances out to about 20 Mpc. The paper explains how to calculate the value of the fluctuations, taking into account various sources of contamination and the effects of finite spatial resolution, and demonstrates, via simulations and CCD images of M32 and N3379, the feasibility and limitations of this technique.

In "The expanding universe and the Big Bang", the development of these theories is described. Fred Hoyle coined the phrase that came to be applied to Lemaître's theory, referring to it as "this *Big Bang* idea" during a BBC Radio broadcast in March 1949. The *Big Bang* is a physical theory that describes how the universe expanded from an initial state of high density and temperature. Various cosmological models of the *Big Bang* attempted to explain the evolution of the observable universe from the earliest known periods through its subsequent large-scale form. They offered an explanation for a broad range of observed phenomena, including the abundance of light elements, the *cosmic microwave background* (CMB) radiation, and large-scale structure. Detailed measurements of the expansion rate of the universe placed the *Big Bang* singularity at an estimated 13.8 billion years ago, which was considered to be the *age of the universe*.

Perlmutter, S., Aldering, G., Goldhaber, G., *et al.* (1998). *Measurements of Omega and Lambda from 42 High-Redshift Supernovae.*, reports on the *Supernova Cosmology Project*, which was started in 1988 to determine the cosmological parameters of the universe using the *magnitude-redshift* relation of Type Ia *supernovae*. All *supernova* peak magnitudes were standardized using a SN Ia *light-curve width-luminosity relation*. It determined that the data are *strongly inconsistent* with a $\Lambda = 0$ *flat cosmology*, the simplest inflationary universe model in which the universe continues to expand at a *constant rate*, for which the best-fit age of the universe relative to the Hubble time was $t_0^{flat} = 14.9_{-1.1}^{+1.4}$ (0.63/h) Gyr. The data indicated that the *cosmological constant* was *non-zero* and *positive*, indicating an *accelerating universe*. [This relation was based on a *relativistic cosmological model*, so is inconsistent with New Physics.]

Lubin, L. M. & Sandage, A. (2001). *The Tolman Surface Brightness Test for the Reality of the Expansion.*, was included because it was one of very few recent "mainstream" articles which claimed to evaluate *Zwicky's "tired-light" theory*, but in fact it did not. Although Allan Sandage was one of the most influential astronomers of the 20th century, he was 75 years old at this time. It was very disappointing. It omitted an expansion factor, then failed to include the *"tired-light" factor* representing the linear loss of energy as the *photons* interact with *electrons* and other matter as they travel through intergalactic space. The evidence for an *expanding universe* which it provided based on *Tolman surface brightness test* was demonstrated by Lerner, E. J., Falomo, R., & Scarpa, R. (2014). *UV surface brightness of galaxies from the local universe to z ~ 5.*, to be in error, and, when calculated correctly, *concluded that far from disproving a non-expanding cosmology, data by Lubin and Sandage agreed very well with predictions for a static Euclidean universe.* However, this paper provides some useful insights on this analysis, and on the deeply ingrained biases at work.

Mamas, D.L. (2010). *An explanation for the cosmological redshift*, is a bizarre attempt, by an unrecognized author based in Florida (with a PhD in physics from UCLA), to reformulate *Zwicky's "tired-light" theory* in terms of a simplistic pseudo-classical notion of a *photon* as an electromagnetic wave which causes a *free electron* to oscillate and reradiate. It is included here together with Mamas, D.L. (2015). *Cosmological redshift model now experimentally confirmed.*, as it provides useful background information in the absence of any mainstream paper. It also provides calculations that suggest that the *cosmological redshift*, based on the quantum mechanical formulation of *Zwicky's "tired-light" theory* and estimates of the density of *free electrons* in intergalactic space, support a *static* rather than an *expanding universe*, on which the *Big Bang* origin of the universe is based. It incorrectly assumes, as Zwicky stated, that *Compton scattering* suffers from the blurring of images.

"The effect of a strong magnetic field on a dielectric" describes how Mamas, D.L. (2010) is reminiscent of a recent explanation by the author of the *Faraday effect*, in which, when addressed in terms of *non-relativistic quantum electrodynamics* (or New Physics) "the *electrons* in the *dielectric* under the influence of the strong *magnetic field* line up according to their *spin*. The motion thus effected will be *circular*; and rotating *electric charges* create a *magnetic field*. So, the circularly moving *electrons* will create their own *magnetic field* in addition to the external *magnetic field* on them.

Shaoa, M-H. (2013). *The energy loss of photons and cosmological redshift*, is another rather bizarre article in which Zwicky's *"tired-light"* theory is presented in terms of the simplistic interpretation of the wave-particle nature of the *photon* and classical electromagnetic theory. This is rather surprising as it was published by Xinjiang Astronomical Observatory, Key Laboratory of Radio Astronomy, People's Republic of China, though not by a mainstream theoretical physics institution. It is included for the same reasons as Mamas, D.L. (2010).

Lerner, E. J., Falomo, R., & Scarpa, R. (2014). *UV surface brightness of galaxies from the local universe to z ~ 5.*, is an important article that demonstrates errors in previous claims that have been made that the Tolman test provides compelling evidence against a *static model* for the universe. This was reconsidered by adopting a *static Euclidean universe* (SEU) with *a linear Hubble relation at all z* (which is not the *standard Einstein–de Sitter model*), resulting in a relation between *flux* and *luminosity* that is virtually indistinguishable from the one used for ΛCDM models. Based on the analysis of the UV SB of luminous disk galaxies from HUDF and GALEX datasets, reaching from the local universe to z ~ 5, it was shown that the *surface brightness* (SB) *remains constant as expected in a static universe*. In particular, it shows that the conclusions in Lubin & Sandage *are not supported by the data* for two main reasons: (1) for the *static scenario*, Lubin and Sandage set the distance to d = (c/H_0) ln(1 + z), *which is valid only for the Einstein-de Sitter static case*. This is not the cosmology being tested where the *Hubble relation* is hypothesized to be d = cz/H_0 at all *redshift*. The conversion factors to transform arc seconds to pc in the *non-expanding model* are therefore different; (2) *the local sample includes only first rank cluster galaxies, while the high-z sample includes about 20 normal galaxies in each of three different clusters*. This means that their distant galaxies are on average smaller and less luminous, and therefore are not directly comparable to local ones because of the well-known *absolute magnitude-SB relation*. *It concludes that far from disproving a non-expanding cosmology, data by Lubin and Sandage agree very well with predictions for a static Euclidean universe.*

The current volume concludes that *there was no Big Bang, the universe is not just 13.8 billion years old*, but *is indefinitely old, and in a steady state, not expanding.*

11   Cosmic Microwave Background Radiation January 9, 2025

The *cosmological redshift of light* is generally assumed to be due to the *Doppler effect* on light resulting from the *expansion of the universe* following the *Big Bang*. However, my previous book, [November 2024), *The Cosmological Redshift of Light.*], concluded that a better explanation is provided by Fritz Zwicky's 1929 tired-light theory as elaborated by Arthur Compton in 1923. Zwicky's *"tired-light" theory*, attributes the *linear redshift with distance from the observer* to the *loss of energy by photons*, and consequent increase in *wavelength*, resulting from *interactions between the photons and intervening electrons or matter* whilst travelling through *intergalactic* space. Compton had confirmed that scattering of electrons by photons results in a redshift due to the transfer of energy from the photons to the scattered electrons.

Cosmic microwave background (CMB) radiation is black body microwave radiation that fills all space which was first discovered in 1965 by American radio astronomers Arno Penzias and Robert Wilson. [See Penzias, A. A., & Wilson, R. W. (1965). *A Measurement of Excess Antenna Temperature at 4080 Mc/s.*] With a standard optical telescope, the background space between stars and galaxies is almost completely dark. However, a sufficiently sensitive radio telescope detects a faint background glow that is almost uniform and is not associated with any star, galaxy, or other object. This glow is strongest in the *microwave* region of the radio spectrum.

According to the *Big Bang theory*, which as noted above, was first proposed in 1931 by Georges Lemaitre, a Roman Catholic priest, after American astronomer, Edwin Hubble reported in 1929 a linear relationship between the distance and redshift of galaxies, the *cosmic microwave background* (CMB) is *relic radiation* from the *Big Bang*. However, as measurements of the CMB improved, contradictions with this theory's predictions began to emerge and this theory began to fall apart, and alternative explanations began to emerge. The analysis in this volume suggests that the *cosmic microwave background* (CMB) *radiation resulted from clouds of ionized plasma from thermonuclear reactions in stars colliding with clouds of intergalactic dust.*

The first section describes the *cosmic microwave background* (CMB) *according to the Big Bang theory*. (This is based on an article in *Wikipedia* titled "Cosmic Microwave Background" which assumes *the Big Bang theory* without noting this explicitly.)

This introduction is followed by the beginnings of an alternative theory. Hannes Alfvén recognized that a *plasma* pervades the universe and applied *plasma physics* to cosmic radiation [Alfvén, H. (1937). *Cosmic Radiation as an Intra-galactic Phenomenon.*), for which he received the 1970 Nobel Prize in Physics.; to magnetic storms [(Alfvén, H. (1937)

A theory of magnetic storms and of the aurorae.)]; to electromagnetic-hydrodynamic waves [(Alfvén, H. (1942). *Existence of electromagnetic-hydrodynamic waves.*)]; and to the origin of the solar system, in his book [Alfvén, H. (1954). *On the origin of the solar system.*".].

[*Plasma* is one of four fundamental states of *matter* (the other three being *solid*, *liquid*, and *gas*), characterized by the *presence of a significant portion of charged particles in any combination of ions or electrons.* It is the most abundant form of ordinary *matter* in the universe, mostly in *stars* (including the Sun), but also dominating the rarefied *intracluster medium* and *intergalactic medium.*]

This is followed chronologically by the first of a series of articles [Erickson, W. C. (1957). *A Mechanism of Non-Thermal Radio-Noise Origin.*] describing a mechanism by which radio-frequency radiation similar to that found in the *cosmic microwave background* (CMB) can be generated by collisions between a high-velocity plasma cloud and clouds of interstellar grains.

Penzias & Wilson (1965) [*A Measurement of Excess Antenna Temperature at 4080 Mc/s,*], as noted above, showed that measurements of the effective *zenith noise temperature* yielded a value about 3.5° K higher than expected. A possible explanation in a companion letter [Dicke, R. H., Peebles, P. J. E., Roll, P. G., & Wilkinson, D. T. (1965). *Cosmic Black-body Radiation.*], raised conflicts in cosmology with *Einstein's field equations* such that *neither the origin of matter nor of the universe could be understood.*

This article proposed three main attempts to deal with these problems: (1) the assumption of continuous creation which avoids the singularity by postulating a universe expanding for all time and a continuous but slow creation of new matter in the universe; (2) the assumption that the creation of new matter is intimately related to the existence of the singularity, and that the resolution of both paradoxes may be found in a proper quantum mechanical treatment of Einstein's field equations; and (3) the assumption that *the singularity results from a mathematical over-idealization, the requirement of strict isotropy or uniformity,* and that it would not occur in the real world. If this third premise is accepted tentatively as a working hypothesis, it carries with it a possible resolution of the second paradox, for the *matter* we see about us now may represent the same *baryon* content [protons and neutrons] of the previous expansion of a closed universe, oscillating for all time.

In the next article Eric J. Lerner, an expert in plasma physics, presents his *plasma model of the origin of the light elements and the microwave background* as an alternative to the *Big Bang* [Lerner, E. J. (1988). *Plasma model of microwave background and primordial*

140

elements: an alternative to the big bang.]. The model assumes that helium, deuterium and the microwave background were all generated by massive stars in the early stages of galaxy formation. The microwave background is scattered and isotropized by multi-GeV electrons trapped in the *jets* emitted by active galactic nuclei. The model produces reasonable amounts of heavy elements, accurately predicts the gamma-ray background intensity and spectrum, and explains the statistics of quasars, and compact and extended radio sources.

Lerner's model (GOLE) was published in an expanded form in his 569-page book [Lerner, E. J. (1992). *The Big Bang Never Happened—A Reassessment of the Galactic Origin of Light Elements* (GOLE) *Hypothesis and its Implications.*], and in a chapter in Arp, H. C., Keys, C. R., Rudnicki, K. (eds) *Progress in New Cosmologies*, in which he claimed the *Big Bang theory* is presently without any observational support. By contrast, he claimed *plasma cosmology theories* have provided explanations of the *light element abundances*, the *origin of large-scale structure* and the *cosmic microwave background* that accord with observation. He noted that it was time to abandon the *Big Bang* and seek other explanations of the Hubble relationship. [See Underwood, T. G. (November, 2024), *Cosmological Redshift of Light*.]

We then return, chronologically, to a study of the radio emission from rotating, charged dust grains immersed in the ionized gas constituting the thick, Hα-emitting disk of many spiral galaxies. [Ferrara, A., & Dettmar, R. -J. (1994). *Radio-emitting dust in the free electron layer of spiral galaxies: Testing the disk/halo interface*.] Grains were found to have substantial radio emission peaked at a cutoff frequency in the range 10-100 GHz, depending on the grain size distribution and on the efficiency of the radiative damping of the grain rotation.

The next article [Lundin, R., & Marklund, G. (1995). *Plasma vortex structures and the evolution of the solar system—the legacy of Hannes Alfvén*.] returns to *cosmic plasma physics* to address the present knowledge of plasma vortex structures and how that impacts on cosmogony, the evolution of the solar system, as it was conceived by Hannes Alfvén in 1954.

This is followed by a response by Lerner [Lerner, E. J. (1995). *Intergalactic Radio Absorption and the COBE Data*.] to new COBE satellite data on *cosmic background radiation* (CBR) isotropy and spectrum, which were considered to be explicable only in the context of the *Big Bang theory* and to be confirmation of that theory. Lerner claimed that the new data could also be explained by an alternative, non-Big Bang model which hypothesizes *an intergalactic radio-absorbing and scattering medium* and gives a better to fit to the spectrum observations than does a pure blackbody.

The next two articles, by Draine and Lazarian, return to *spinning dust grains*. The first [Draine, B. T., & Lazarian, A. (1998). *Diffuse Galactic Emission from Spinning Dust Grains*.] claims that spinning interstellar dust grains produce detectable rotational emission in the 10–100 GHz frequency range of which the emission spectrum can account for the "anomalous" galactic background component which correlates with 100 μm thermal emission from dust. The second [Draine, B. T., & Lazarian, A. (1998). *Electric Dipole Radiation from Spinning Dust Grains*.] discusses the rotational excitation of small interstellar grains and the resulting electric dipole radiation from spinning dust. It addresses the excitation and damping of grain rotation by collisions with neutrals, collisions with ions, "plasma drag", emission of infrared radiation, emission of electric dipole radiation, photoelectric emission, and formation of H_2 on the grain surface, claiming that spinning dust grains could explain much, and perhaps all, of the 14-50 GHz background component.

This is followed in 2003 by an interesting internet article by Dr. Edward L. Wright [Wright, E. L. (2003.) *Errors in the "The Big Bang Never Happened"*.] commenting on Lerner's 1992 book "*The Big Bang Never Happened*" and Lerner's 1995 article, *Intergalactic Radio Absorption and the COBE Data*. Wright is an American astrophysicist and cosmologist who received his PhD (Astronomy in 1976) in high-altitude rocket measurement of *cosmic microwave background* (CMB) *radiation* from Harvard University and had subsequently worked on space missions including the Cosmic Background Explorer (COBE), Wide-field Infrared Survey Explorer (WISE), and Wilkinson Microwave Anisotropy Probe (WMAP) projects.

In 2008, Lerner responded to Wright's comments. [Lerner, E. J. (2008) *Dr. Wright is Wrong -- a reply to Ned Wright's "Errors in The Big Bang Never Happened"*.]

The next article [Fixsen, D. J. (2009). *The Temperature of the Cosmic Microwave Background*.] reports on the latest refinement of the measurement of the *cosmic microwave background* (CMB) *temperature*, independently recalibrating the Far InfraRed Absolute Spectrophotometer (FIRAS) data using the Wilkinson Microwave Anisotropy Probe data. This confirmed the temperature of intergalactic space, to be 2.72548 ± 0.00057 K.

Ali-Hamoud, Y. (2012). *Spinning dust radiation: a review of the theory*. reviews the current status of theoretical modeling of electric dipole radiation from *spinning dust grains*, concluding that robustness of theoretical predictions now seems mostly limited by the uncertainties regarding the grains themselves, namely their abundance, dipole moments, size and shape distribution.

Kroupa, P., Pawlowski, M., & Milgrom, M. (2012). *The failures of the standard model of cosmology require a new paradigm*. notes that cosmological models that invoke warm or cold *dark matter* cannot explain observed regularities in the properties of dwarf galaxies, their highly anisotropic spatial distributions, nor the correlation between observed mass discrepancies and acceleration. It introduces an alternative in the form of *Modified Newtonian Dynamics* (MOND), a classical dynamics theory which explains the mass discrepancies in galactic systems, and in the universe at large, without invoking 'dark' entities. It concludes that the remaining challenges for MOND are (i) explaining fully the observed mass discrepancies in galaxy clusters, and (ii) the development of a relativistic theory of MOND that will satisfactorily account for cosmology.

This is followed by an article [Schwarz D. J., Copi, C. J., Huterer, D., & Starkman, G. D. (2016). *CMB anomalies after Planck.*] reporting on several unexpected features that have been observed in the microwave sky at large angular scales, both by WMAP and by the recent *Planck* satellite. Among those features was a lack of both variance and correlation on the largest angular scales, alignment of the lowest multipole moments with one another and with the motion and geometry of the solar system, a hemispherical power asymmetry or dipolar power modulation, a preference for odd parity modes and an unexpectedly large cold spot in the Southern hemisphere. It notes that despite numerous detailed investigations, we still lack a clear understanding of these large-scale features, which seem to imply a violation of statistical isotropy and scale invariance of inflationary perturbations.

The next article, [Ćirković, M. M., & Perović, S. (2018). *Alternative explanations of the cosmic microwave background: A historical and an epistemological perspective.*], traces historically various non-conventional explanations for the origin of the *cosmic microwave background* (CMB).

Will Handley [(2021). (*Curvature tension: evidence for a closed universe.*] notes that without *cosmic microwave background* (CMB) *lensing* or *Baryon acoustic oscillations* (BAO), *Planck* 2018 has a moderate preference for *closed universes*, with Bayesian betting odds of over 50:1 against a *flat universe*, and over 2000:1 against an *open universe*.

Abdalla, E., *et al.* [(2022). *Cosmology intertwined: A review of the particle physics, astrophysics, and cosmology associated with the cosmological tensions and anomalies*] notes that the standard Λ Cold Dark Matter (ΛCDM) cosmological model provides a good description of a wide range of astrophysical and cosmological data, but there are a few big open questions that make the standard model look like an approximation to a more realistic scenario yet to be found.

Finally, the most recent online update by Lerner, E. J. [(October, 2022). *The Big Bang Never Happened—A Reassessment of the Galactic Origin of Light Elements (GOLE) Hypothesis and its Implications.*] notes that the growing list of failed predictions of the *inflationary LCDM theory* is a widely-recognized crisis in cosmology; it is therefore timely to re-examine if the *Big Bang hypothesis* (BBH), which underlies the dominant cosmological model, is valid. The core of that hypothesis is that the universe began with a short period of extremely high temperature and density. Such a hot, dense epoch produces light elements by fusion reactions. But the actual published predictions of the *Big Bang Nucleosynthesis* (BBN) theory of light element production have increasingly diverged from *observations*. The predictions for both *lithium and helium abundance* now differ by many standard deviations from *observations*, a situation that is worsening at an accelerating pace. Only *deuterium* predictions have remained in agreement with *observation*. In contrast, the published predictions of *the alternative hypothesis, that all light elements were created by thermonuclear and cosmic ray processes in young galaxies, have been repeatedly confirmed by observations. This paper reassesses the galactic origin of light element (GOLE) hypothesis in light of new calculations and recent observations.* The GOLE predictions remain in good agreement with all relevant elemental abundance data sets and are contradicted by none. *Only a single new assumption of EM energy loss with distance is required.*

[This is now provided by Zwicky's tired-light theory as elaborated by Compton. See Underwood, T. G. (November, 2024). *Cosmological Redshift of Light.*]

Nor are any of the quantitative predictions of BBH for the CMB in accord with observations, while *Galactic Origin of Light Elements* (GOLE) hypothesis provides an alternative explanation for the CMB that requires none of the BBH's hypothetical entities, such as *dark matter* or *dark energy*. BBH predictions are contradicted by 16 different data sets while GOLE predictions are contradicted by none. *The solution to the crisis in cosmology is to abandon the Big Bang hypothesis.*

The current volume concludes that *there was no Big Bang, the universe is not just 13.8 billion years old*, but *is indefinitely old, and in a steady state, not expanding.*

www.ingramcontent.com/pod-product-compliance
Lightning Source LLC
Chambersburg PA
CBHW061326190326
41458CB00011B/3909